ALONE

IN THE

FIRE

THE FIRST ALARM

by

Stephen Davis

CHOICE
PUBLICATIONS

For my late Papa, Laomedon Glenn Stephens, my grandfather who shared his life and Christian faith with me as a boy, young man, husband, and father.

ACKNOWLEDGEMENTS

In my life, I have been driven by many people who have inspired, motivated, and coached me to be a better version of myself. To those individuals, thank you. You gave me a heading in which to travel.

Early in life, I learned to be a leader. My parents, Danny and Linda Davis, were there to show me good Christian family values, love, support, and accountability. They provided a caring home for my two sisters, my brother, and me. Without their support through the challenges I have faced in my life, I would not be the man I am today. Thank you for your love through the years.

To the men and women of the fire service, I want you to know that this book was written to empower and encourage all of you to speak up when something is wrong. Too often we have sat at the dinner table and complained without action. But when we all stood up together and acted after the mandates were put into place, we saved the careers of many Orange County employees. The Mayor and his Commissioners wanted to take our jobs if we did not comply. Instead, all they got was a written reprimand. That's because four hundred of us were brave enough to Stand Up and Speak Out. Imagine if we continued that stance today. Do Not Be Complacent!

To those close friends who have stuck with me through this entire process... Thank You. It doesn't go unnoticed. You have done more than I can acknowledge in a single book.

To my wife, Veronika, and my sons, Caiden and Kai, this journey over the last couple years has not been easy for us, but you have been my sole purpose to fight harder and harder every day. You have been the reason I get up and push myself to every limit there is in this world. The courage I have comes directly from inside me to give you the best life I can. While I was left Alone in the Fire in my profession, I have always had you three with me, keeping the fire burning inside hotter and brighter each day.

Lastly, to my Lord and Savior, Jesus Christ. You have opened many doors for me to walk through and closed many that were no longer needed in my life. I continue to be a servant under Your grace and protection.

Thank you.

CONTENTS

PREFACE

"You shall have no other gods before me," - The First Commandment.

"First, do no harm" - The Hippocratic Oath.

"Do to others what you want them to do to you," - The Golden Rule.

"Innocent until proven guilty," - The Fifth Amendment to the U.S. Constitution.

"No one is above the law," - The Rule of Law.

"We hold these truths to be self-evident, that all men are created equal, that they are endowed by their Creator with certain unalienable Rights, that among these are Life, Liberty and the pursuit of Happiness." The Declaration of Independence.

Contrary to popular opinion, these statements are not cliches. No, rather, they are the truths upon which the foundation of our world is built.

Unfortunately, cracks are forming in the once strong moral wall of our society. Sadly, there are those out there that are more than happy to nitpick at that wall's mortar, creating larger and larger fissures. This book, *Alone in the Fire* by former Firefighter, Paramedic and Fire Chief Officer Stephen Davis endeavors to shine a light on those fissures and cracks. He accomplishes just that!

I first met Stephen Davis many years ago in Fire School at Central Florida Fire Academy. He was one of thirty-three new cadets to be hired following a cattle-call testing of nearly 1,400 applicants. The first day on the fire training ground, I was told by the fire instructors that cadet Davis would be our squad leader and we had better follow his lead.

"What was it that caused the instructors to single him out as the leader of squad five?" I wondered to myself. The answer became apparent when we were preparing to enter a human-sized oven of a steel box called The Trailer, where we were going to experience one of the deadliest situations a firefighter could see... Flashover.

Flashover is when all present fuels meet their ignition point simultaneously, reaching upwards of 750 degrees Fahrenheit. Cadet Davis instinctively perceived and effortlessly calmed our fears by checking our equipment one by one and reminded the four of us of what we had been told. "Do not stand up in this extreme theater of flame or the fire dragon will swallow you whole. Stay low and we'll be fine. This is gonna' be awesome," he stated with a quiet confidence that he probably learned as a medic in the US Army. That was the moment I knew that cadet Davis would one day be addressed as Chief Officer, Stephen Davis.

The Covid 19 pandemic brought with it a firestorm of death and destruction. Fear spread like flame lengths that scorched people's lives like so many trees in a forest. As burnt trees began to fall, a phenomenon we call "snags" in the fire service, they hit piles of ashes on the forest floor. Amongst the fallout, we are now beginning to find that our moral and social fabric is burning and piling up as well. However, underneath the piles, there is still a golden moral compass. This book by Chief Officer Stephen Davis does two very important things at the same time. One, it's a call to action, and two, it points to the place where that compass can be found again.

So, as one first responder to another, (yes, you are a first responder as well even if not by job description), I encourage you to dig into the pages of this book, *Alone in the Fire.* And, if you dig long and thoughtfully enough, you may just find that the true moral compasses will be resting upon an old leather bound and scorched, but fully intact Word of God which, by the way, is where Chief Officer Davis ultimately finds his guidance as a leader.

- Paul Lawrence, Pastor, and retired Fire Lieutenant of Summit Fire & EMS, Colorado

INTRODUCTION

A Monday morning, October 18th, 2021, I was contacted by Michael Howe, Assistant Chief of Training, and lead investigating officer for my Pre-Determination Hearing. He called me and asked me if I would come into headquarters. I had been relieved of duty for thirteen days. I was in trouble. I had 'failed to follow a direct order.' Little did I know what was in store for me...

After the call, I remember thinking, 'Finally, the results. I can head in, and more than likely be suspended. I'd been with the Orange County Fire Rescue Department in Florida for fifteen and a half years; I'm valuable to them; they'll just suspend me. I can handle that.' I did not believe I was facing a termination.

INTRODUCTION

I made a call to my wife, Veronika; I told her that I was heading into headquarters to find out what was going to happen to me, not knowing that this was going to be the end of my career in Orange County. I called my fire union. They had no idea that I had even been contacted.

I headed into headquarters, complying with their mask-wearing mandate as a show of good faith to my superiors. Those of you that know me know, I don't wear a mask. I never have. But that day, I complied with the directive.

I walked into the conference room; I sat down. The Termination Board consisted of the following individuals: Battalion Chief Shannon Teamer; Assistant Chief Martis Mack; the Local 2057 Vice President Paul Riccardi; and Assistant Chief Michael Howe, who read the order as we all sat around the conference table.

I remember them looking at me with what appeared to be confusion. I don't want to say sadness because they knew what was to be my fate. They knew that I was to be terminated. I could feel it in the air, like the tension before an execution. They handed me my termination papers; I flip straight to the page where it says you're terminated, and Michael began reading it to me verbatim, a blank tone in his voice hidden and muffled by his mask.

I ripped my paper mask off with disgust and threw it on the table. So many thoughts went through my head: From thinking I was going to be suspended swift to a sense of righteous anger and wounded outrage and all I wanted to do was to throw and break things.

I had given so much to Orange County Fire Rescue in fifteen and a half years. I had achieved so many things in such a short amount of time. I was one of the youngest individuals to earn the rank of Battalion Chief, and I did it in ten years. I was working towards my master's degree in Fire Science. I was just a month shy of having that. I was on the Assistant Chief's list, next in line to that promotion. I had started the dive program for the County and had revamped the paramedic program. I was involved in every aspect of Orange County. I felt betrayed; I felt hurt; I was sad, and I was angry.

They terminated me for misconduct. They terminated me for insubordination; they said that I had violated a direct order.

But no, I had refused an *unlawful order.*

INTRODUCTION

Orange County policy states that if you have any reason to believe that an order is unlawful, whether it be a state statute, a federal law, or a county ordinance, you have the duty to refuse that order.

Orange County Fire Rescue Rule 33 and Rule 35 states that, "Employees shall not knowingly issue any order that is in violation of laws, statutes, ordinances, rules and regulations, or SOPs. (R.33) and "No employee is expected to, or shall obey any order, that he/she knows to be contrary to federal or state law or County ordinance." (see Appendix, pg. 222)

I knew that this order was in fact unlawful, and for its defiance, I was betrayed by my department. I was terminated. This job was my only stream of income. This was the only way that I put food on the table. I had given everything to the Fire Service. Everything that I had done, two decades of work, everything, all surrounded Orange County Fire.

Everything, and it was gone in a second.

What was this unlawful order?

I was given a list of over 100 names and ordered to issue written reprimands to personnel that had been

identified as not complying with the Covid-19 vaccination mandate.

The list of individuals that was presented to me was incorrect. There were people on the list who had religious exemptions; there were people on the list who had medical exemptions; there were people on the list who in fact had the vaccination and had their vaccination card, and yet I was told to write them up. There were people on the list that never even knew of the order because they were out of country for four months and didn't return until the mandate was in effect.

If you turn now to the Appendix of this book, pg. 199, you will see all of this in primary sourced documentation, gathered from my extensive public records requests. What was my boss' response?

"Issue the written reprimands, and let them grieve it later", Kimberly Buffkin, Assistant Chief of Orange County Fire Resue, told me.

Yet the order was unlawful.

The order was a violation of people's civil rights. The order violated state statutes, Federal Law, even international Law, the Nuremberg code! Getting rid of me, firing an employee with my authority, responsibility, and integrity, was setting an intimidating example. If I, literally a poster boy for the County, could be gone, so could anyone. Who would dare stand up to such a ruthless bureaucracy?

How do we stand up for our convictions in a world that is hostile to them?

Masculinity, speaking up for what's right, self-sacrifice; these are some of my values, and it was my adherence to these values that led me into the fire and *led me through the fire.*

In this book, you will hear my story, *Alone in the Fire*, and how I stood up against what was wrong. In this book, I'll tell you what it means to be a man in a world where men are denigrated; what it means to speak up for what's right; how to have the toughness to face the consequences.

That's what leaders do. In this book, I'll teach you how *you are a leader!* Whatever your responsibilities are, you have an obligation to fulfill them and to rise to a life of excellence and integrity.

It was my responsibility as a senior Chief Officer to be a voice of reason, to be that person that expresses the concerns of the subordinates whom I oversaw, whom I was responsible for, whom I commanded. It was my responsibility as a Battalion Chief to speak up for them, but it was also my responsibility to maintain the integrity of my department.

In this book, you will see my trial by fire, and how that fire burned so bright, it created a global movement of people standing up against tyranny. My trial by fire burned me into a better man. My trial by fire was the crucible wherein was made a parenting movement committed to raising the next generation of leaders not followers, **The Raising Alphas Project.**

As I leave my firefighting career behind, as I move on to bigger and better things, I leave with you my story of the man who stood up against corruption, who said something when he saw something, and wasn't afraid of the consequences. And even though my career had been taken, I was also given an opportunity to do something bigger and better.

Yes, sometimes we go through hard times, but that's because through those hard times, you can get to somewhere better. That's how God challenges us.

INTRODUCTION

Remember this, God gives the toughest battles to His strongest warriors.

This is my battle.

Chapter 1

The First Alarm

Firefighter Davis at his first fire

"What you stood up for; you truly saved lives for what you did!" John Haskett, Orange County Fire Rescue Battalion Chief (Ret.)

Emergencies, that can encounter us any day and any moment, are our calls to action. Some of us run away from it, some of us run to it. I was called to action very young. I was six years old when I heard the first alarm.

I was at a friend's house, his name was Darren; he had an older brother, Dustin, and a younger sister, Alicia. We had just gotten out of the pool and were all dressed, and Alicia - too close to the edge - had suddenly fallen into the pool.

There was no adult out there with us. I remember Alicia falling into the pool and going down and not coming up. She was about four years old. At first, we all froze. One precious second ticked by.

Then, immediately, on instinct, I sprint inside the house to Mrs. Sue, Alicia's mother. Quickly, I tell her what's happened: Alicia's fallen into the pool. She's not coming back up... I very vividly remember the look on Sue's face and how she, in her Sunday dress, runs outside and leaps into the pool and grabs Alicia from the bottom and pulls her back up to the surface as quickly as life itself.

I remember her emotional state. Her scream. But Alicia was OK. She was just fine. She had been rescued, just in time. Mrs. Sue came over to me and gave me, not a hug, but an embrace. She was crying and she was thanking me, and she was holding me and holding Alicia.

I was the call to action. I saw something and I said something, and it saved someone's life.

That's what this book is about, *Alone in the Fire*. Working for Orange County for 15 and 1/2 years I saw something, I said something, and my intent was, like when I was six years old, to save a life. My intent was to correct what was wrong, to save an agency from embarrassment; I saw something, and I said something.

At age 6, my first exposure to the life-saving profession (which is more than just the uniform) was the catalyst that took me into a life of looking after others, of being someone who would stand up for what is right, who sees things, recognizes them, assesses the situation, and makes the decisions of a leader.

I didn't know it then, but at age 6, God had a plan for me. He has a plan for you too. I believe that even though I have made plans myself throughout my entire life, it was His plan that I was executing. His plan was for me to be in a profession that I've held for over 25 years, and today, He is still opening doors to walk through and closing doors to pass over.

Which door is He opening in your life? Which door do you know He has closed?

A Christian Household

I grew up in church and in a very religious home. We went to church every Sunday; we went to Sunday school; we wore our best Sunday dress. We went to the Sunday night service and the Wednesday night service too. I relied on my faith. I relied on God and Jesus always.

I come from a very Christian lineage. My grandfather, Glenn Stephens, was a pastor. His pastor joke, which he used throughout his life, was that he was a pastor, and then he was a teacher, and then he went to jail. Well, when he went to jail, he was actually working for the jail system as a chaplain.

He helped inmates come to find the Lord and set their paths straight.

Glenn, my Papa, was very strong in the faith, and I remember him sending me scripture verses in the mail, I still have them today, when I was going through the military. When I was in the Army's basic training, he sent me a letter every single day. A proverb or a scripture would always end the letter. I still cherish every word.

Our faith carries us through our fates. Through all the things I've gone through, especially with Orange County, my faith has gotten stronger. When you get put into dark times, we all look to something, for someone, strong. Take it from a firefighter who would know, people always turn to God in their emergencies.

Have you ever heard the expression, Man plans, and God laughs?

I've often said over the last couple years that my plan was to be in Orange County for 30 years. But this wasn't God's plan. I was wrong. He closed the door and opened another and pushed me right through it!

Shortly after my termination, I was picked up by another county. Lake County Fire Rescue hired me within 13 days. Immediately, I had a new plan in mind. My plan was now, not to finish my career in Orange County, but in Lake County.

God laughs when you make plans!

God laughs, and I say that jokingly, but I say that because, yes, He wants you to have direction; yes, He wants you to plan for things, but at the end of the day, *your little plans are always shaped into His greater plan.*

The Lake County door was soon also closed, more about that in my next book. I was with Lake Country Fire Rescue for 1.5 years as the Division Chief of Training and Special Operations. Then I retired. I left because God had bigger plans for me. He opened my mind and the way I look at things long before Orange County terminated me. A bigger purpose was in store for me, and in order to reach that peak, I had to leave what I loved and cherished behind forever.

I left for what was most important, my family, and my mission of raising up true leaders, what would later become Raising Alphas.

I believe that my work with Lake County was for me to gain closure with my sudden divorce from Orange County. So, my plan became His, perhaps it was His all along. Now I'm in a league of my own. And I believe that no one can keep up with me and where I'm headed now.

I could not have done any of this or pursued the opportunities that I'd pursued if I had not been raised and lived steeped in my Faith. I am raising my own sons the same, so that they will be ready for their own battles.

Raise Your Children with Faith

Why should we raise our children in a house that is full of faith?

Everyone reading this today can look back on some time of their life when they have gone through great difficulties. How did you handle that? Did you search for your God, did you fall on your religion? Were you strong in your faith?

When emergencies alarm, even the faithless find faith. But don't wait till you most need it. *Nurture your faith now, so that when you go through the difficult time, you and your family will be ready and already strong.*

One of the programs that I was raised with, that steeped me in my faith and prepared me for leadership, was the Royal Rangers program. The Royal Rangers are similar to the Boy Scouts, well, when the Boy Scouts were the *boy* Scouts.

The Rangers is where young men learn how to do things like camping, fishing, hunting, building tents, making campfires, having fellowship, being a team player. For each lesson, we could earn a badge.

The Rangers taught me to be self-reliant. They taught me leadership and teamwork. They taught me about faith and hard work and how hard work leads to success.

At the beginning of this chapter, I told you about the near drowning of Alicia. Well, her father Tim was my platoon leader in the Royal Rangers. After his daughter was pulled out of the pool, he presented me, in front of all the other Rangers, a Red Badge of Courage.

If you've ever read the war novel by the great American author, Stephen Crane, *The Red Badge of Courage*, you'd know the significance of a medal like this. In the novel, a red 'badge' was a wound you received in battle. It was the corporeal and proud proof of one's courage. It showed that you were willing to bleed for your band of brothers.

Sometimes, we prove our courage with wounds. Remember, alarms are calls to action. Sometimes, those courageous actions we take *takes a toll* on us.

Of course, my badge at six was nothing like the severity in Crane's novel, but it was still an honor for me to earn it. It showed that I helped to save a life. It showed that I was able to react and act upon the occasion with no hesitation.

I saw something; I said something.

Alicia's dad presented me that red badge of courage to show that this young man showed boldness, without fear, without hesitation, with leadership. What he gave me was a simple card, with a sticker on it; but it meant everything and to me it was everything. It was like the seed from which my life grew. I still have it today. Tim presented it to me with emotion. Such a little thing stood for so much: his daughter's life.

How many lives did I save by standing up against Orange County? How many lives have heard my story, and chosen not to get the vaccination, chosen to stand up against the many tyrannies that arose at that time, and the tyrannies that will arise tomorrow?

My story has been heard around the state, the country, and the world. The governor of Florida, Ron de Santis, publicly recognized me. I received national attention. I was featured on multiple national news reports. Movements against governmental tyranny in

California, in Los Angeles, movements in New York city, from Australia to Indiana, all started after my termination.

Those men and women are still fighting the good fight today against all the varieties of corruption and injustice. These are the battles that make us.

You have the opportunity every day to earn your own badge of courage. Whether you're a leader in your own home, whether a leader at work, or even just a follower right now, you're somebody that can influence the people around you in a positive direction. Are you gonna answer your call to action?

Aiming High

I always wanted to be a fireman. At 14 years old, as a step on the path to that goal, I became a lifeguard. It was my first job. I worked for a water park called *Wet N Wild* in Orlando Florida. I met firemen there, and I would excitedly ask them, how do I become a fireman?

I was aiming high. After graduation, I joined the Army. I requested to join the firefighter track, but they made me a medic. That early medical experience would be invaluable to me in my later career. There was my plan, then God's better plan, remember?

The Army taught me discipline, excellence, and integrity. I earned the expert field medic badge. I went to flight medic school and became a flight medic out in California. In the military, leadership was one of our core values. You can turn the word leadership into the following acronym:

L oyalty

D uty

R espect

S elfless service

H onor

I ntegrity

P ersonal courage

Those seven words, commitment to them is what has driven me for years; acting out that acronym is what has inspired this book.

I could have soared even higher, gone into the Green Beret program, but I had other goals. After my time in the Army, I immediately got picked up with Orange County Fire Rescue.

In the Army, I got to be all I could be, but now, finally, finally, my dream became my reality. I was determined to be the best fireman ever. I wanted to be the top, the best of the best.

To end this chapter, I want to tell you about one of the calls I had while in the Fire Service. Just like when I was six, the alarm of emergency sounded for a child drowning...

The Drowning

I've run on many drownings. The one that stands out the most was the young boy who had fallen into a puddle of rainwater in a drained-out backyard pool. The puddle had been there for months, green, leafy, and gross, maybe 4-feet of water. The boy had fallen in, and his parents didn't know where he had gone.

They found him floating in it. He was quickly pulled out by the first responding deputy. There was no time left to lose.

Meanwhile, we were all outside playing basketball, my whole crew, at the busiest station in

Orange County, Station 51. Ironically, it's the last station I ever worked at there. We were used to our alarm, a loud tone that would go off, and an automated voice that would say 'Medical' if it was a medical call, as it so often was. Then we would switch on our radios to 'Tach 5' to get further updates, its automatic voice droning, 'cardiac arrest', 'chest pains', 'stroke', 'auto accident', etc.

But this time, an experienced dispatcher jumped on the radio for us. She was our alarm, and she spoke with urgency:

"Station 51, engine rescue 51, you have a drowning, a pediatric drowning!"

Her voice set a fire on our feet and we sprinted, all of us, two to the rescue truck, three to the fire engine, away from the basketball court to the fire station, each of us answering the call to action.

I, as the medic, was in the rescue truck with my partner who was driving, Ryan Stanley. Police officers of the Sheriff's Office were tactically stopping traffic. You may have seen something like this happen on a Hollywood TV show. But that doesn't happen on a regular basis in the real world. It's usually an engine (the engine is a truck with pumper to supply water to a house fire) and a rescue (the ambulance unit) trying to fight through traffic, honking its horn, trying to push through the clogged traffic of Orlando.

But on that day, it felt like 1,000 deputies came out to halt traffic. I'd never heard the city so still. It seemed like on every road every car was out of our way. They had moved everybody. They had stopped traffic to a standstill. They made a way for us. We seemed to get to the scene in the blink of an eye; I hoped that it was not one blink too late.

The officer first on scene had already started CPR. I remember me and my other partner, Ryan Allen, getting the child. He brought the kid out to me. I took the boy into my arms, maybe 5 years old. I put him in the back of the rescue unit, and we immediately took off, heading to the nearest ER. I am working on the boy, intubating him while also maintaining high quality CPR. When I say, 'intubating', I am essentially putting a tube down the child's throat so I can breathe for him. I breathed for him. We arrived at Arnold Palmer hospital.

The boy was taken inside, where he would lie in a bed, and live for three more days.

He wasn't responsive.

It was one of those tough calls. You may have seen some of these Hollywood shows, how the medics always go back to the hospital to follow up on their patients; that doesn't usually happen either. In fact, it very, very rarely happens... Medics are always on the go – there are just too many patients.

But this time, I followed up with the patient and his family. That's how I learned that the child had passed away after three days. You cannot save everyone. Alicia was saved. This boy, ultimately, was not. But I still answered the call. And in this case, because of all our heroic efforts, that boy's parents were able to be in the presence of their child and say goodbye to him with dignity, in a place where he was laid gently in a warm bed, breathing softly on life support.

I've had to answer many drowning calls since I was six years old, but it's this one that stands out the most for me because I gave that family three more days. I gave that family an opportunity to say goodbye one last time, knowing that we, the men and the women who have been sworn in to answer the alarms of emergency, did everything we could.

I wish I could have given them more. At the end of the day, I am grateful that I was able to provide them with a little bit of comfort in the midst of their battle.

Remember what I said, *God gives the toughest battles to the strongest warriors.* What will your battle be? Will you answer your alarm? Are you ready to?

Chapter 2

The Readiness

Squad 5: Rick, Matt, Mike, Paul, & Steve

"Practice' gotta be harder than the games..."
- John Calipari, Head Coach, University of Kentucky
Men's Basketball Team

At twenty-five years old, I finally made it to a job that I'd been wanting to work my entire life. I'd been through the training; I'd been through weeks and weeks of preparation. My first structure fire came about a year later. I was ready.

I was on the Rescue at Station 70. The alarm came in. We were to be first on scene for a two-story apartment fire. I get on the truck, and my heart is racing. My mind is clear. I am completely focused, confident in my training, confident in my comrades, confident in myself.

The fire was raging on the 2nd floor like a furnace. There was no report of anyone inside the structure. We still did a full search of the building. My Battalion Chief said simply, "let's go to work.", and that's what we did.

I'm holding the nozzle, my partner feeding me hose, and we storm in, breaking down the front door with our fire axe and Halligan bar. The heat and darkness were incredible. My partner and I stepped into it without hesitation.

I could feel the heat, the heavy heat, the hot temperature, the black smoke pushing through the air, moving like a walking shadow. We fight it. I advance the line through the apartment and found the seed of the fire, where it was the hottest, where it had started, where it was getting its energy from, the very heart of the blaze.

I opened the nozzle bale. We used a fog nozzle, but I had it set to a straight stream for this attack and went to work. After just a few moments of this fire *fighting*, I could feel a temperature change. I was having an effect. Our Chief Officer would be able to see a 'conversion' in the smoke. That's when an aggressive, turbulent smoke, escaping hot and fast through the broken windows and any other opening, begins to slow down and to ease up, turning from black to light grey and white.

As a young fireman, I never paid attention to these supervisory details. I was just excited about getting after my first fire, going head-to-head with 'the beast'. Every bit of training, all those long hours and days in the sun working on liquid propane props... now, I finally got to see, feel, and fight the real thing firsthand.

I got to be in the moment of pure danger, and I loved it!

But that moment was the culmination of a well-honed readiness. How did we, as firefighters, equip ourselves with knowledge? We must know how to 'read' a fire before we 'fight' a fire. So, how do we read a fire?

The first thing we consider with apartment fires is that the buildings are made of wood framing. So, they are what we call 'light weight trust systems'. If this fire gets up into the attic, we need to be concerned about the roof coming down on us. We have to be concerned about potential victims inside. We have to be concerned about water supply. Sometimes, we're only working with 750 gallons; that's what we carry on our fire engines.

We have to know how to break down doors safely and efficiently. Our search and rescue is a deliberate process. We conduct what's called a Right-Handed Search. We follow the right-handed wall, through the entire structure, like navigating a maze blind. This is so we all know which way to go and can essentially follow or find another firefighter in the darkness of the smoke.

We stay in communication; we stay in constant visual/physical contact with our partner. We do not leave them alone in the shadow of the smoke. In some cases, when you're going into a highly heated environment, you cannot see the hand in front of your face. In those moments, you have to rely on your training, your team, and ultimately, yourself, your own inner strength and fortitude.

Training to be a Firefighter

My fire academy was in Orlando, just down the street from Station 51. I remember being named one of the squad leaders, and my squad and I were very close. We did a lot of stuff together. We had lunch together, we trained together, and we hung out after the academy together. It was a close camaraderie, like brothers.

Those individuals are still friends to this day. One of those team members, Paul Lawrence, his son became my book publisher, Joseph Lawrence with his Choice Publications. I met him when he was a young child. I remember going into his father's house and meeting his three boys. Then, years and years later, Joseph is now a friend.

Fire School was training from the morning you woke up to the moment you got home. Now, I was coming out of the US Army, so I had already been prepped for years for this kind of high-intensity environment. But for a lot of guys this required a mindset adjustment, or you'd wash out and get dropped.

You must have the discipline to do the physical and mental training. Each day, we did physical training and then school, so it was constantly doing stuff. You're in school all day, and you've already been working out in the mornings. PT, classroom, then fire training. Day after day. The grind-set had to become your mindset.

Fire training was breaking down doors, loading fire hose, tying knots, climbing stairs, pulling dummies out of staged car wrecks, or crawling through a narrow cement cylinder small enough to pin your arms to the floor, but just big enough to breathe in. We did these 25-yard crawls in full bunker gear – overcoat, mask, air tank and all. If we ran out of air before we got through the cylinder, we could suffocate.

Confined spaces... It takes a lot to stay calm, to stay focused, in confined spaces. These were the intense tests that firefighters had to overcome. If you couldn't handle these moments in a controlled environment, how could you manage to keep yourself calm when it's chaotic and out of control?

All our tests were superheated. Our training took place during the hot summer in Florida. Let me tell you, there's nothing like putting on bunker gear in the Florida heat and working with an even hotter fire. Our bunker gear would increase our body temperatures by at least 10°. If it's 95° out, then you're working at 105°. We were fighting fires in a sauna while wearing an oven over our skin.

But I can tell you this, fire school was fun. You heard me right: It was fun! Our attitudes are everything. If we can get enthusiastic about our challenges, if we can enjoy the grind, then we can turn those challenges into victories, those grinds into gains.

The Rookie

I graduated in September of 2007. Fire Academy was about 3 months, and immediately after, we went into Fire Orientation. I was already a certified paramedic for the state of Florida, so after orientation, I was to go on shift. I felt great, ready to go into the service as a rookie.

I'd come from the military, so I knew what being a rookie was like. You have to play the game. You must be the first one up in the mornings; you have to empty the dishwasher, make the coffee, wash the truck, make sure the station is clean. You must be a servant.

You have to ensure that you're the last one to go to sleep, if you have an opportunity at a station to sleep. In a busy city, there are fire stations that never sleep. As a rookie, you'll be the last one to dress down on your uniform. You don't get to wear your T-shirt until late in the evening. Your shoes must always shine. Your pants must always look ironed. You must always look sharp.

Hazing was a thing in the firehouse. It was a way to bring you into your second family. It was a way to see how you can handle the heat and keep cool. Hazing gets a bad name today because people take it too far sometimes. But it's also a tradition to bond strangers together who at any point could lose their lives.

The "Brotherhood" has been lost in the fire service. This can only be blamed on cowardice: officers who climbed the ranks and forgot where they came from. This can also be attributed to firefighters looking after themselves as well. But more on this in my next book, where I talk about the Brotherhood in the Fire Service.

I remember being hazed as a rookie, as a paramedic, as lieutenant, as a Battalion Chief, and as a

Division Chief. It happened during my entire career. It's what keeps up respect because it reminds us that we're all still human. Nobody is above (or below) a well-meaning prank.

As a Chief Officer, pranks don't go as far, but I've been hit in the face with a cake a few times. I've had my shirts frozen because I left them in the Gen. Pop. Area. I've had my shirts hung from the ceiling of the Engine Bay about 35ft up with the rope just out of reach, right before I was leaving to go home.

Did I get pissy about this and start yelling?

No!

I laughed and joked and schemed for my own next attack. It's what bonds firefighters together. It's the way to relieve the pressure of a tank ready to blow...

During my rookie year, I wanted to be completely submerged in the culture of the fire service. I wanted to take it all in. I had been waiting so long for this moment, and so my goals were to learn everything I could. I remember my engineer, 30 years on, was trying to show me how to pump the engine. It felt like I was becoming fluent in a foreign language. He wanted to take the time to teach me, so I could learn his profession.

I learned so much, like firefighter survival, like how to rappel on a high-angle rope. I applied for as many advanced level classes as I could. *Each challenge was an opportunity for positive change.*

You see, Fire School is the minimum standard. Some guys rely on those minimum standards their entire careers. They never take on the challenge of an advanced level course. This is a bad decision for firefighters because you get complacent, and when you get complacent, you become dangerous, not only to yourself, but to your team and the people you're there to rescue.

In contrast, advanced level courses, by their very nature as challenges, will keep a firefighter sharp, ready, and highly qualified. Advanced level courses for firefighters are things like Hazardous Materials, Confined Space Technician, Trench Rescue, Rope Technician, Structural Collapse, and Rescue Diver. Advanced courses for the EMS side of the job are classes like EKG recognition, advanced airway recognition, and critical-care classes.

All these courses are to prepare a firefighter to excel at any task they're called upon to perform. I knew early on in my career that I had to challenge myself. I had to be the best fireman. And if I was going to be the best fireman, I had to know the highest amount of knowledge.

This is what leaders do. Leaders engage with the challenges that followers leave un-attempted.

Initiation Rituals

After their first structure fire at Orange County, a firefighter will have the privilege of having their shirt cut off by his/her fellow crew members. It's one of many of the ceremonial aspects of being in the Fire Service.

I remember getting my shirt cut off in the parking lot of a hospital. I had just transported one of my fellow firemen to the hospital due to heat exhaustion. Just as I took my bunker coat off, a couple guys that were on the fire had also transported, and I remember them taking my shirt and shredding it with a razor and scissors.

So now I'm walking around the parking lot with a half-cut and shredded shirt on... I throw my bunker jacket back on, showing my whole chest underneath, and I don't have another shirt. I remember getting back on the Rescue, and immediately as we're cleared – bam – another call went out. We ended up getting a medical call. So I'm tending to the patient practically shirtless, and then we're transporting back to the hospital, still with only half a shirt.

Jokes like, "Well, this is exactly how I want a fireman to show up to my house!" was overheard more than a few times...

I've kept that shirt for over 17 years now. It is a token of my rite of passage. It memorializes my lifelong life dream I've had since I was a young boy. That shredded shirt says that I have become somebody that goes *into* a burning building, that I'm running in as everyone else is running *out*. And the fact that my teammates cut the shirt showed that I was a part of a team of these brave men and women who run into fires, not away from fires.

The Brotherhood

My career started with a group of guys in Squad 5 during the Fire Academy. We all had an inseparable relationship. Matt, Paul, Dan, Mike, Ryan, Rick. We created a lifelong bond of brotherhood.

I will go deeper into the idea of brotherhood in the Fire Service and beyond in my next book. Stay tuned for that! But I will say a little bit about it as a preview of the things I will discuss there.

These bonds are important because when you're going into an untenable and unsafe environment, when you're confronting an environment that can kill you, that is immediately dangerous to life, where the toxicity in the atmosphere is so concentrated that it can kill you with just a few breathes, you need to be able to rely on your buddies absolutely.

We firefighters have such strong bonds because we are going to war. We are in battle every day we step on the truck. At any point in that day, we may not be coming home at all. So, at any point in that day, we are looking after the person next to us, our partner, our teammates.

We are training together all the time so that when the moment comes, we can operate as one body, not several. Being able to anticipate, work in sync with each other, and find our own rhythm, allows us to move efficiently, safely, effectively. That's how we save lives.

This lesson can apply to your life too.

Surround yourself with others who want to rise to higher levels. In terms of career, in terms of faith, in terms of leadership, surround yourself with people who push you to be better. When you take on greater challenges with others, you rise higher than you would alone.

The Pandemic Begins

When COVID-19 came into the picture, it created a pandemic not just of disease, but of fear, of mistrust, of hostility. It was a pandemic of division. So many people had differences of opinion about how we were to respond to the alarm, the emergency.

"Wear a mask."

"Don't wear a mask." "

"Don't come near me."

"Stay at home."

"Stay at work."

I can tell you that in the fire service, if you had the right crew, and the right bonds, those guys didn't allow the pandemic to come between them. They stayed strong. But in a department like Orange County, which has such a large variety of men and women from so many different backgrounds, cultures, religions, walks of life, political beliefs, you were going to have a divisive moment. There're over 1,500 employees in the Fire Department. To think that the pandemic, and the actions taken by the County in response to the pandemic, was *not* going to create division is insane. COVID-19 created such a severe strain on the fire service, on that brotherhood, that you could see people taking their own lives...

I will speak more about those tragedies in a later chapter.

Because of COVID-19, and the County's response to it, people began to hate one another. What was being done wrong? How could this possibly have been so poorly managed? Where were our true leaders?

Leadership in O.C. Fire Rescue

What were my last years at the County like in comparison to when I first joined?

In my early years, I can tell you that the fire service was about preparation, about getting better; it was about becoming a stronger fireman, whether in special operations, whether as paramedic, whether it was improving and promoting, *progress* was the spirit of the place.

As I grew in my career, I promoted up the chain of command. As I promoted, I had more and more responsibilities with different objectives to take on, different accomplishments to meet, but through all that change, one thing stayed the same: I always wanted to improve those around me. That's why I took on promotions.

When I became a Lieutenant, I tried to improve the training at my firehouse. I tried to get my crew to think differently about the typical challenges we faced. I was also being improved as much as I was improving.

At Station 73, Delsie, Rob, and Julian were mentors. The four of us had a great relationship. They taught me a lot as a young supervisor. Delsie was my engineer, and I had once been her rookie firefighter. I became Robb's Battalion Chief. We're still close friends. Julian also remains another good friend of mine. Those bonds I made also made me.

My earlier years of fire service were a great time, but later on, as I began to promote, I became more and more exposed to some of the corruption and the poor politics. I became exposed to some of the ways that senior leadership handled things.

This was a burden. In some ways, I wish I had not promoted. Sometimes, I wish I was still just a regular firefighter riding backwards on a fire truck. Because ignorance is bliss.

But I had more in me than that. God had a purpose for me; he had more things for me to do. So, what would I tell myself years later as a rookie?

I would encourage myself to continue to strive hard, not to let stresses get to me, because as a firefighter paramedic, I had a lot of stressors. I was at one of the busiest firehouses in the state. I had patient care. I was constantly on the go. I would tell myself to sit back and enjoy the present more, not to promote so fast. Stress increased with responsibilities.

High levels of stress and cortisol lead to a lot of heart attacks, strokes, and other complications. Firefighters face these challenges. I would tell myself, and I would tell this to all who are facing stress, to have good health, to eat well, to be physically fit.

Would I have told myself about the fight I was to face with Orange County over the Covid mandates?

I would tell myself to trust my instinct always; to trust that God has a purpose for me, and His plan is greater than mine.

I would tell myself, "You will face the worst betrayal you've ever felt in your life. What you will go through will be hotter than any fire you have ever fought in your life. It will be more intense than any emergency scene you have provided care for, more challenging than any large-scale incident you have commanded. Stay true to yourself and trust that God is there with you." I would tell you the same thing.

Chapter 3

The Plagues

OC Firefighters tested hundreds every day.

"See Something, Say Something"

March of 2020 was the beginning of a worldwide pandemic. It was an unprecedented event in the United States. I remember lying in bed with my wife just after we got back from Germany that January of 2020. She was showing me images and videos of things happening in China. She was showing me how tyrannical governments broke into homes and grabbed people, arresting them for not wearing masks. I remember laying there and telling her, "That will never happen here in the United States. We're a free country with an iron-clad constitution."

For those that don't know, my wife is European, so while she's lived here in the United States going on two-decades now, she still has questions about how things happen here. I, being the man of the house, being the leader of my home, it is my responsibility to reassure her. I tell her that everything's going to be ok.

March of 2020 rolls in; the pandemic is here. In the fire service, we were ordered to reduce the amount of personnel we sent on calls. We used to send four firefighters on an engine, and two on a rescue truck. That was reduced to only two individuals on a rescue.

All we could tell those two was, for any type of call, wear masks, wear gloves, wear eye protection, cover your entire body with a plastic suit. Have you ever worn one of these hazardous material suits? They can get pretty warm. You're responding to calls in the middle of the summer, into and out of homes without A/C, carrying the equipment that normally four other people would be helping you carry...

I remember one guy telling me he had changed his shirts four times, stuck those in the laundry, and asked if there was a way to get more for the shift. This was all before noon that day. He had already borrowed one from a crew member. Yea, it was that hot.

In our early stages of 2020, the rescue crews were by themselves the majority of the time. They were the ones taking people to the hospital during the scare tactics the media was peddling – we were told that this was the deadliest virus that ever plagued mankind. I was the Battalion Chief, a leader in the third largest agency in the state of Florida, and I have no answers for these men and women serving on the front lines of the disease.

In early 2020, we'd locked down businesses. People were losing their jobs. You're not able to go to the gym. You can't go to the grocery store and walk down the aisle in freedom. There're arrows pointing you which way to go, as if that's going to solve this disease. 6 feet. Stand back 6 feet.

THE PLAGUES

Just think of everything that was going on during this time: people were fighting over toilet paper; people were not able to pay their rent or mortgages; people were not able to attend church.

I had young children at the time, my oldest was in kindergarten. He was not able to have a real kindergarten graduation. I wasn't even able to walk him to class on his first day. They forced him to wear a face-shield. I refused to put a mask on him, so they forced him to wear a face-shield... A 5-year-old... A 5-year-old who was very active every day. That face shield came home with abrasions on it, obstructing his view. Is that something that I should be concerned about with the school board? The school, two years later, sent home to me a failed eye exam. Is this linked to the face shield they forced on my child? Should I hold them accountable for his poor eyesight? If I don't, who will? Who will speak out for our children?

2020 was a challenge most of us had never experienced before.

When you start talking about firemen who have been in and out of horrific calls their entire career; when you have put firemen in situations of traumatic injuries, tragedy, sadness, there's something that weighs on them. Now you're telling them that they're gonna have to work more with less help, more fear with less certainty, what does that do to a psyche?

"Hey chief, is this something that's going to be contagious?"

"Is this something that I'm gonna take home to my family?"

"Is my family gonna die if I go home?"

These are questions firefighters were asking me.

"What happens if I take someone to the hospital, they have covid, and then I contract it, and I take it home to my wife and children? What do I do if they become sick? Do I go to the hospital? Do I stay at home?"

We at Orange County were told to stay at home. I submitted to that at first. I played their game for a couple weeks, and then I realized that it was stupid. I stopped doing it.

I remember going to the park in Orange County, and I remember it being completely taped up with caution tape. Nobody else was in the park, and I'm told that I cannot play with my kids in the park?

Screw that!

I ripped down the caution tape. I went ahead and allowed my children to be free in this free country. Orange County's response to the fearful pandemic was not like mine.

Losses of Life

Let's think about everything that a firefighter has to go through in a career, and now imagine you're isolating them; you're not allowing them to work out; you can't go out and have good food; we're promoting alcohol; we're promoting fast food restaurants; we're forbidding church. What effect does that have?

Right before the turn of the year to 2021, Orange County Fire Rescue lost two firefighters to suicide, two senior officers who had been in the fire service for a long time and had retired. In early 2021, two more firefighters take their own lives.

Now we're talking about four Orange County firefighters, in a span of five months, who had taken their life by suicide with a firearm.

Why is this happening?

I don't know.

However, one could argue that with two of them being 30-years-plus in the service, seeing as much pain and suffering over those years as that, and then being locked in their homes, isolated, cut off from human contact... How much does that turn up the pressure? Were they just too worn down from it all?

But the other two firemen who took their lives were young. They were in their early 30s. They were early on in their careers, only a couple of years on the line...

Why did this happen?

In that same time frame, a good friend of mine and comrade in the service, Eric, passed away from his battle with cancer.

Eric was a good friend. In the mornings, he would come into my bunk room and wake me up. He'd have his little white mocha coffee, sit down in my recliner that sat next to my bed and yell "Hey, get up! It's late, it's 6 in the morning!" Eric was the A Shift Lieutenant, and he always got there early.

I remember the night that he discovered the tumor in his brain.

THE PLAGUES

I was sitting right next to him at the fire station. We were at the dining table. His Battalion Chief, who would become the assistant chief who would terminate me a couple of years later, was not on duty... Eric had a seizure right in front of me. He had taken a bite of chicken, and he seized up at the head of the table.

I pick him up. I put him on the floor. I pull the chicken out of his mouth, rip his shirt off, call over his crew. We all started working him right there with our high-quality medical care. I remember us putting him in the back of the rescue. I remember calling the assistant chief that was on duty that evening. I remember going to the hospital.

I remember Eric waking up.

I remember him looking at me. I remember him looking at the Fire Chief who had shown up. I remember Eric looking at me again and giving me the dirtiest look... He did not care for that Fire Chief at all... I'll talk more about him another time.

So, in about a 5-month period in Orange County, we had five servicemen lose their lives. It was a very difficult time in Orange County. I remember going to the memorial services for a couple of them. It wasn't an easy thing. Eric's was the hardest.

Eric was my friend. Eric was my hazmat pocket guide, my guy that I would call if I was on a hazmat scene. 'Hey Eric, this is what I got, what do you think?' His wife was also a good friend of mine. She and I had worked as paramedics at a local firehouse. We'd worked together for years.

I remember being the lead vehicle as we carried Eric's body to the memorial. We started at his house. I carried his helmet to his fire truck. We carried his remains on that fire truck all the way from Lake County to the center part of Orange County, our sirens and lights on, not moving to an emergency, but *moved* in the wake of one.

That was the hardest time of the year.

PTSD & Firefighters

PTSD in the Fire Service hasn't been recognized until recently. In the last few years, it's become more & more prevalent in conversation. It's becoming more and more discussed, and that's because of the toll it's taken.

When you talk about PTSD in the military, trauma happens so quickly. It happens in an instant, and you lose your friends, or a part of yourself. In the fire service, you're seeing horrific things over a long period of time, not all within a single moment.

So, in some cases, it takes years to actually identify the trauma. And firefighters have ways of getting around it, ways of suppressing it, of being a man, being tough. We laugh. We joke. We have a very rude sense of humor. We drink. We do all these things to cope and to suppress those feelings.

As a Chief Officer, I gave signs of PTSD my strictest attention. I've recognized how severe a scene could be, how horrific, and judging by debriefings, and hearing some of the responses of our men and women, I would make the call to let certain individuals take the night off.

For another example, I'd have small groups of maybe 5 to 10 firefighters on my dive team, and I would take the opportunity when they came in and did training to talk for 5-10 minutes. 'Hey, if there is anything that you want to talk about...'

What I thereby wanted to show to these individuals is that we've gone through a lot in the last year together being separated from people. We've suffered from the dividing of our world, of our country, of our families even.

In these small groups I would tell them, "Listen, if you notice something, someone acting different, say something."

The method worked. I remember hearing one of my lieutenants say that an engineer at his firehouse was posting these dark memes, which were not in this guy's nature. I knew this guy for a long time, he and I have been close friends, and I saw it, and I knew something was up.

We saw something. We said something.

As a friend, I reached out to this guy. He was going through some personal issues at home, some trouble recently at work; the house of cards was getting ready to fall. He didn't return my calls. He didn't return my messages. Was he going to be another casualty?

I called his crew members and asked them to look out for him. They did. Those small groups were a life-raft for us. How many lives did we save with them?

This is about you too.

If you have people that are posting things on social media that's uncommon for their personality, say something. If you see something, say something. The last thing we want is another individual to lose their life.

I ran into that firefighter a few months later on a smoke alarm call, and asked if everything was OK. He said that he had indeed been going through a hard time, but thanks in part to our support, was doing better now.

A year after that conversation, another Orange County firefighter committed suicide.

Was this ever going to stop?

Mandates & Corruption

I did not like the mandates. I did not like the lockdowns. I disagreed with it. 'I have to stay in my home, I can't go out'. 'I have to wear a mask.' 'I have to take this shot.'

I think it's ridiculous.

Politically, I think President Trump did what he could with the advisors he had. I think this man led our country in a great way. I think, unfortunately, the swamp is so deep that, in some cases, you just don't know who to trust. You think you have a trusted advisor, someone that's been doing this for 20-30 years, I can trust him. But you can't.

We'll talk about Anthony Fauci. Anthony Fauci is an evil man. If you haven't had an opportunity to read Robert Kennedy Jr.'s book, *Real Anthony Fauci,* you should. Take the time to. It's an excellent, complex read. It reports on everything Fauci has been doing his entire career...

For Governor Ron DeSantis, I can't say enough good things about him. When he heard about my termination, he immediately brought me into his circle. I'll tell you about the day I first met him in a later chapter, but I will say that he did a great job in the state of Florida. He looked past all the nonsense. He played the game for a few weeks, but then he opened everything back up.

It makes no sense that you can't be walking on the beach. It makes no sense that you can't be walking in your neighborhood, playing at the parks. It made no sense.

Governor DeSantis and the Orange County mayor, Jerry Demings, were going toe to toe on this issue. Jerry even alluded to the fact that the governor was using me as a pawn in his scheme. 'A political pawn' is what he called me.

Jerry Demings closed everybody down. He closed the playgrounds. He closed recreational sports. What kind of pride can a man possess to dictate us in this way?

It's sickening.

I remember that year my boys were not able to play soccer because of this pandemic. And yet, the Orange County Convention Center, which brings in millions of dollars for Orange County, was left open.

Let that sink in.

Jerry Demings shut down playgrounds but kept his county's cash-cow wide open.

Orange County was the only government in all Central Florida, in all of the state, that had this response: mandated vaccinations, and locking everybody down.

Because Orange County is the largest in Central Florida, smaller municipalities began to waver, began to follow suit, to give in to tyranny. Isn't that how it goes?

One falls, then another. What could stop this domino effect of liberty's ruin?

All it takes is one solitary voice.

But there's a catch: He who would speak up against tyranny must be prepared to face tyranny's fires.

Fortunately, I was good at fighting fires.

Chapter 4

Alone in the Fire

Battalion Chief Davis saw something, said something – Terminated!

"An organization's worst nightmare is when its most passionate people go silent" - Tim McClure

The day of my termination, March 18th, 2021, I arrived at Orange County Fire Rescue headquarters. I remember greeting a couple firefighters that were sitting at the front desk. I remember having a smile on my face as they greeted me. They didn't know what was going to happen to me. I didn't know either. I stood there for a solid 5 minutes, conversing with them like it was just another day at the firehouse. Then I got summoned in.

As I'm walking down a long hallway, there's another hallway to the right, and there was the termination board, standing and scheming. I continued into the room. I sat down, and I waited patiently.

I was in good spirits because I believed that the decision being made was going to be a suspension. I hadn't had a single infraction up to this point, not a single blot on my record, nothing that would indicate I had to be terminated for what I did. At this point in my career, I was heavily involved with Orange County Fire Rescue. I was a very dedicated member of that agency. I was well educated. I was properly trained. I had advanced training. At no point did I see what was coming for me.

Would I have acted differently on that day if I could sit in that room with that termination board again?

To be honest, there is no way for me to know. To this day, I know I have a strong case. To this day, I know that what I did was right. I saw something wrong, and I said something to a superior to correct it. In no way did I feel that what I did was wrong.

I will tell you this, if I could do it all over, if I could sit in that room with that termination board again,

I would've been a bit more *forward*. I would be more direct, because at that moment I took the high road, at that moment, I *accepted* what they said.

I am not the individual then as I am today. I would have challenged them. What else did I have to lose at that moment? I think I may have had them call security on me! I say that jokingly, but I feel strongly against their corruption, and when you're up against evil, you must challenge it in the most abrasive and direct way possible.

I was betrayed by my department. I was betrayed by individuals that I believed respected me, that I had respect for up until that moment. I believed that at the right moment, the moment when it really mattered, that they would do the right thing. They didn't.

At that moment, I realized I was the only person doing the right thing. At that moment, I was alone in the fire, and I was going to be the one to walk through it.

Staying Positive in our Fires

Remember the acronym I told you about?

75

My core values have stemmed from that acronym, leadership.

LDRSHIP.

Loyalty, Duty, Respect, Selfless-service, Honor, Integrity, and Personal-courage.

So, because I've lived on those core values since I was eighteen years old, it wasn't hard to stay true to those values in the face of peer pressure. I talked about peer pressure a lot when I was a Battalion Chief. I talked about it a lot as a Lieutenant. I talked about it as just a senior fireman. Peer pressure works in two ways, a positive and a negative.

Too many times in the fire service we see negative peer pressure. Too many times we see negative peer pressure in our lives, and it's easy for us to go along with it. This is opposed to doing the right thing, which is *projecting a positive outlook on life.*

Positivity. It's wonderfully contagious. Conversely, there's nothing more discouragingly infectious than an individual who comes into the firehouse with negativity. Negativity. It's a disease.

Too many times I've heard firemen say sarcastically, 'We have the best job in the world!' Too many times I've heard firemen say, after I ask, 'Well how's your day going?' 'Living the dream', is the response. 'Just living the dream...' but as you look into it a little bit more deeply, as you look into the true reality of a firefighter's life, are they really living the dream?

I'm 'living the dream', yet I'll come into fire stations, and I'll complain. I'll be negative. I'll take my time to get to the truck when a call is dispatched. It's just another medical call. Oh, it's just another fire alarm... Too many times in the fire service, we face peer pressure on the negative, infectious side.

What does the opposite sound like?

While a leader in the fire service, I talked about positive peer pressure. I spoke about being that positive influence for the people around you, being a guide for people to see a better light. It's very difficult when you're the only person, or you're in a small minority of people, who believe strongly in positivity. So, what my advice is to everybody reading this book:

Stay steadfastly positive when you have people around you who are negative.

And what if the negativity never changes?

Then, do not be afraid to change your environment. *Change your circle if you're the most successful person in your circle.* Find another circle. Find people who are more successful than you are to be around, who are more positive than you, and feed on their energy, learn from what they have gone through, from the fires they have walked through.

First Reactions to the Termination

I had walked into those doors at headquarters an employee of Orange County. I was about to walk out of them an unemployed husband and father of two young boys. And those two firefighters I spoke with on my way in, now I had to tell them that I had been fired. I told them that I was upset, that I was angry, and that I couldn't even stand to be in that building anymore. I was so angry, I remember going out to my vehicle, sitting there, and immediately making a video of protest and complaint.

Here is a quote from that video:

"This isn't the end. I am going to fight this. This is what happens when people remain silent. This is what happens in this place. And until people start standing up, we will never get through this crap. Everyone will be held accountable for their actions. As a Veteran, I have served this country as a Non-Commissioned Officer... so now they have one pissed off solider. They messed with the wrong dude. Today is the Beginning!"

I never posted that video. I recorded it so I could constantly remind myself of the feeling that I felt after I left that building: the anger, the rage that poured through my blood. I wanted to remember the moment that still drives me today.

After I recorded the video, I told my wife Veronika that I had been fired. She was at work. I remember she left work to come home to be with me. I didn't tell my kids right away.

It's my responsibility as a father, as a husband, to be the person that leads the family, to lead and protect my loved ones from stressors. Everything's on our shoulders. So, while my wife knew, my kids didn't know right away. Later, I simply told them I didn't work for Orange County anymore. I didn't want them to ever think that there was anything wrong. I wanted them to be kids, to live their life.

To stress about things, that's my job. The financial uncertainty that faced me was a very difficult feeling. I've always been a planner, so having finances set aside, having something that I was able to tap into, was invaluable. To this day, I practice planning for financial wealth. Because you don't know when a storm is going to hit you and your family. You need to be ready.

After my firing, several firefighters stood up for me and helped take care of things. Wendy Williams and Jason Wheat. These two individuals came around right away and started a Gift-Send-Go campaign. Michael Molaka started a GoFundMe campaign. All the gifts and expressions of support from family and friends helped a lot. Those campaigns afforded me the opportunity to live my life as if I was still a Battalion Chief in Orange County.

I never missed a payment on my vehicles. I never missed a mortgage payment. Christmas was coming up, and my kids still had a good Christmas. I had already paid for a family trip up to north Georgia and had friends that were meeting us there; I was still able to take that trip. I was still able to maintain quality of life for my family during that time.

I give a lot of thanks to those three individuals, and to all that helped me through this time.

Thirteen days after my termination from Orange County, I had already gotten picked up with another job with Lake County Fire Rescue, with a promotion. I went from terminated to promoted. But I learned a valuable lesson from that experience; I learned how important it is to have multiple streams of income. We'll talk on this real quick.

Basics of Financial Security

Every day, you go into work, you clock in, and you clock out, you get paid for those hours that you work. If you're a firefighter who works a 24-hour shift, you get paid for 24 hours. Here's the thing, as soon as you clock out, *you're no longer getting paid.*

That sounds obvious, but you may not realize that by clocking out you've made a choice, you have chosen not to participate in passive income. There are tons of it out there. I have a passive income through real estate. So, when I clock out of a nine to five job, I'm still getting paid through my passive income streams.

I've also started my own business. Have you looked into opportunities to start a business and make money after you clock out? Passive income is the best kept economic secret in the US. It's the path to financial freedom. What we're taught from the moment we enter the workforce at age 15 or 16 is this: Clock in, clock out, pay your taxes. That's it.

We're not taught how to *invest.*

Too many firefighters are invested in one thing, their retirement. I hear it all the time: 'I can't wait to retire!' I will tell you, if you are only thinking about that one investment, and you clock in and clock out, and have no other streams of income, you will never be wealthy. Once you reach that retirement age, your expenses will be more than what your retirement brings in. How many people do you see today who are of retirement age still working?

I can go on for an entire book on how to achieve these things, things that I've learned in the last several years by doing my own research through reading, through moving in different circles. There's so many avenues to achieving financial freedom.

How do you know when you're financially free?

You're not financially free until your passive income is more than your expenses every single month.

It's that simple.

The 4-Step Grievance Process

Orange County doubled and tripled down on their unjust punishment of me. It's not a simple process to fire a Battalion Chief in the Fire Service. There are four steps of appeal to it (Step 1,2,3, & Arbitration). This is called the process of grievances. My arbitration phase got abruptly canceled by my union later on in the year without notice. No one ever contacted me, or reached out to see if I was doing ok. My own union, Local 2057, hung me out to dry.

If you're terminated like I was, the grievance process goes directly to a higher ranking official. For me, it was the deputy Fire Chief of Orange County. He makes the ultimate decision. The decision can then progress, going above him to a board, consisting of usually the head of the department, the head of the labor union, and the head of another department.

I was going through my appeal process; would I be exonerated? I'll tell you, but first, I want to give you a reminder on why I was terminated:

I was terminated because I put my name on a paper saying we are not going to allow Orange County to discipline its employees for not complying with the Covid-19 vaccination mandate.

I was the only chief officer who said something after I saw something. The list was wrong. The list of names, names that I was ordered to discipline, had individuals on it with medical exemptions, that could not take the vaccination because it would harm them. The list consisted of individuals who had religious exemptions. The list consisted of individuals who had already gotten the vaccination and had a card and had complied with the order. The list consisted of individuals who were not even in the country at the time of the initial mandate through to its deadline.

These individuals were on the list, and I was told by my assistant chief, Kim Buffkin, to write them up and 'let them grieve it later'.

I was told to do it without question. It didn't matter that I believed this to be an improper and unlawful order. It didn't matter that I believed what they were doing was wrong. It didn't matter that it violated my own religious beliefs, that it violated everything that I stood for. I did not comply with this tyranny of the government, because it was wrong.

Those were some of the facts surrounding my termination.

I remember asking Orange County off the record to allow me to resign, to remove my termination and compensate me for that time, less than $15,000. I had already been picked up with another department; I was ready to walk away and say, publicly, that I'd had a great career at Orange County, that I loved it, that I did nothing wrong. If they'd allowed me to change my termination to a resignation, I would have left this all behind me. No harm, no foul.

Instead, they were tripling down. No, we're not going to accept a settlement. No, we're not going to allow you to resign. No, we're not going to compensate you for anything. We're not going to acknowledge all the facts that you've presented. We're not going to acknowledge everything that we've swept under the rug...

We're going to keep the termination on your file. We're going to insist that everything you did was serious misconduct and insubordination and you're going to remain terminated for it.

I had found out their sins, their corruption, and I was going to be punished for it. What was their corruption? I'm going to tell you all about it.

The Corruption I Found Out

At the end of February, I had just turned forty. I believed that for my 40th birthday, I would be able to win my step 3 grievance and move past all this. I believed that they would give me my job back, so I could ultimately resign with honor and stay where I was in Lake County.

The public safety director denied my grievance. He was the third party. The labor union had voted for me. The county had voted against me. The third party was the public safety director, Danny Banks. I've got more to talk about with Danny Banks.

Danny Banks stated to me in my fire battalion truck a few weeks before my termination that he did not agree with the mayor's vaccine mandate decision, and that no firefighter should be terminated for it. He said that to me in my truck. I was with Danny all day before I was terminated. I believed that he was an individual who had values.

Later, when he denied my grievance, he stated that the reason for the denial was that not one person on the disciplinary list was written up inappropriately, that despite all the hundreds of names on the list, not one was written up improperly.

When Danny Banks denied my Step 3 grievance, I was left with feelings of anger and betrayal. I thought he would be an individual to stand up against corruption. Instead, he appeared to me to be a liar, full of weakness and cowardice.

But the corruption goes deeper than that, and there's a reason that my lawsuit is still an ongoing lawsuit as I write this today.

After I pulled public records, I found that Kimberly Bufkin had hidden evidence in my Step 2 and my Step 3 grievances. She hid evidence in my PDH (Pre-Determination Hearing) and I found those emails. I found documents that had been tossed. I addressed this with Danny Banks before I filed my lawsuit.

I found documents that were unsigned documents, documents that don't exist, documents that were hidden from me in my Step 2 and my Step 3 grievances. I found written reprimands that were originally signed and filed, but then later pulled and shredded. These are official government documents, Public Records that were buried, but I had them. I even found emails from HR stating they received religious exemptions from the list of names I argued for.

Once I found these documents, I emailed Danny Banks. I sent him everything. I showed him where official records had never been recorded in personnel files, where an official HR signature had never been signed (where it is filed officially). I found later on that HR signed these documents after I questioned them... They were signed in March of 2022, six months after they were issued, five months after the Governor signed legislation making this illegal in the state. I told Danny that he was in a position of authority, and therefore one of responsibility, to do what was right.

I provided him with everything.

Nothing more came of it from him.

Attorneys got involved on their side. Orange County refused to communicate with me. At that moment, I had no choice. That's how I decided to file a lawsuit against Orange County for my wrongful termination.

I discovered corruption in Orange County Fire Rescue Agency. I found that there were senior chief officers hiding documents to do one thing, to double and triple down on my case and to railroad me.

I had done nothing wrong. I was a steward for their department. I was the face of that department. I had a good reputation with that department and all the individuals there. People knew who I was, and yet they got rid of me because of a political game.

The mayor of Orange County called me a political pawn. He was so wrong about me.

I am a firefighter. I save peoples' lives. I run into fires. And even though I was alone in this fire, I was not afraid.

Not long after my termination, I got a call from the office of the Governor of Florida.

Chapter 5

Shadrach, Meshach, and Abendego

Davis speaks with Governor DeSantis at the Protect Florida event.

**"I think the firing was totally inappropriate. I think it's
ridiculous. And these are people who have served our
communities, and they're being tossed aside. I mean, it is
illegal to do it, and there's going to be a response."
– Florida Governor, Ron DeSantis**

My termination quickly became a public event,
and that stemmed from many of the firemen who were
in support of my decision to not discipline them on the
mandate. Many of the firemen that supported me were
not ones that were being disciplined in the first place.
They were just people who backed me based on my

stance and my beliefs. Jose Cotti, Jason Wheat, Wendy Williams, Oscar Negron, all showed up in public support. There were several others, Jorge Hernandez, Maria Bernard, Mike Junge, Billy Cezalien, Steve Sherrill, and Chris Newsome. Attention to me grew and grew. It was like a snowball rolling down a hill, gaining speed, gaining mass, ready for an impact.

I didn't have any type of social media at the time, so people were looking to see how to get a hold of me, and Jason Wheat acted as my agent in a way. Everything went through Jason; he would receive phone calls, emails, texts, etc. Pretty soon, the Governor's office heard about what had happened to me and wanted to get ahold of me.

I said to Jason, "I don't want to do any local story until I meet with the Governor. I want my first opportunity to go public to be with the Governor."

It was important that I show everyone just how big a deal my termination was. Because it wasn't just about me. The Governor, Ron DeSantis, was in support of ending the lockdowns. He was not forcing mandates. He was not forcing masks.

He was the perfect individual to come out in support of me, and so I waited. My first public appearance after my termination was with Governor DeSantis down at a small airport in the Tampa area. It was me and two other guests who were to speak with the Governor on stage. One was a nurse from up north, and the other was an airline employee.

I got to shake his hand. He gave me his Governor's coin. My first impression of him was that he was very well spoken. He had a big smile on his face. He told us, "What they did to you guys was wrong. And these governments and agencies need to be held accountable for it."

Not long after meeting with him, he would sign Florida bill HB-1B, stating, "I told Floridians that we would protect their jobs and today we made that the law. Nobody should lose their job due to heavy-handed COVID mandates, and we had a responsibility to protect the livelihoods of the people of Florida. I'm thankful to the Florida Legislature for joining me in standing up for freedom."[1]

---

[1] Staff. (2021, November 18). Governor Ron DeSantis Signs Legislation to Protect Florida Jobs. Florida Governor Ron DeSantis.
https://www.flgov.com/2021/11/18/governor-ron-desantis-signs-legislation-to-protect-florida-jobs/

It would now be illegal to fire an employee for refusing to take a Covid-19 vaccine.

My termination had ignited the fire of his legislation against governmental tyranny. Little did I know at the time, another fire was being lit, one of a global movement for masculinity, liberty, and courage: The Raising Alphas Project.

In the Spotlight

I had to silence my phone because I was getting so many messages and phone calls. I was on the phone all day conversing with different people. All the local channels wanted to have my story. I was invited to the Lincoln Dinner in late October. I was given VIP treatment. I was given VIP passes. I got to shake hands and take pictures with senator Ted Cruz.

The Lincoln dinner was held on Oct 22nd, 2021, at the Rosen Centre Hotel. This was where I was catapulted into the political arena. It was the moment I realized I may be going into another profession.

I started to meet politicians all over the state of Florida. I met people running for Senate, for Congress, state representatives, local politicians, people that were running for mayor, people that were running for commissioners, state attorneys, the list just goes on and on. The head of the Orange County Republican Party, Charles Hart, had brought me up on stage at that dinner. I spoke for about 5 minutes.

Everyone could tell that I was very nervous. But I was grateful to be there. I had nothing prepared. I didn't even know that they were going to invite me up to speak!

After my speech, as soon as I walked back to my table, I was approached by a litany of attorneys, radio hosts, talk show hosts. I was thrown right into the furnace of attention, politics, and big money. Everyone was coming up to my table. Everyone wanted pictures. Everyone wanted something. I was invited to luncheons. I was invited to dinners. I was invited to speak at their campaigns. I was asked to endorse them.

That night I realized the power of being public with my message.

I started prepping things. I invited Orlando Channel 9's Shannon Butler to cover me while I dropped off my gear at Orange County. So, on the day I had to turn in my gear, she showed up with camera in hand. She filmed me taking it out of my pickup and walking inside. She interviewed me for about 15-20 minutes afterwards. Then Channel 35 got ahold of the story. I spoke with Amy Kaufeldt, and she did a big special on me. She interviewed me for about an hour.

I got to be on many national news too. I was interviewed with Attorney General Ashley Moody on *Fox and Friends* with Ainsley Earhardt, Steve Doocy, & Brian Kilmeade. I was on *Newsmax* with Rob Schmitt. I was on *The Dana Loesch Show*. I've listened to Dana Loesch since forever. I was a big fan! Many public figures were giving me national attention that I had never thought I would get. I spoke in front of the Senate to push for Governor DeSantis' law.

The one person that interviewed me that I probably got the biggest attention from, and why it's kind of pivotal to this day, and explains why I do what I do with my Raising Alphas Project, why I have my own show, was because of *Behind the Shield*, a podcast hosted by James Geering.

He and I were firemen together in Orange County, and he ended up leaving Orange County for a local department. He decided to go into podcasting on the side. He started the *Behind the Shield* podcast. It's a great show that "brings the greatest minds in physical and mental wellness to the first responders, military and medical professionals of the world." You can check it out at: **jamesgeering.com**

He had me on his show to tell my story. His show gets international attention, and this is probably the show that really ignited a lot of movements, because as soon as I was on James' show, I started getting phone calls from guys in Australia.

I got in touch with Steve, Dan, Jase, and Josh, Australian firemen, saying, "I love your message. I love what you've done. You know we're dealing with a lot of problems over here in Australia, where everything is censored, and we aren't able to talk about these things."

Steve has a wife who's a physician, and his wife was ridiculed because she did not want to enforce the vaccination. She didn't believe in it. She was taking a risk to speak up against the medical authorities for her own opinions, just like I had spoken up against my own authorities.

I was being publicized all over the nation, and now the world. With these firemen from Australia, we ended up putting together a coalition where we would communicate regularly about the global conflict between liberty and tyranny.

For the 2-3 months between Orange County and Lake County (and Orange County took their sweet time in getting Lake my paperwork – nothing retaliatory about that!), I was very involved in this kind of media engagement. Our Australian-American coalition grew to around 250 people.

Soon my story was being heard everywhere, from New York to California, from Florida to Australia, from Indiana to India. Everywhere that there were issues with governmental tyranny (and isn't that everywhere nowadays?), my story was being heard, the story of the chief that was terminated because he stood up for his crew, the chief that stood up to tyranny and was fired for it.

Those few months changed my life forever. I could not have done what I did without all those individuals standing up and supporting each other and joining the movement. I was in the fire, but I was no longer alone.

Shadrach, Meshach, and Abednego

Growing up in a Christian environment, I was always familiar with the story of Shadrach, Meshach, and Abednego. These were three men who were told to idolize a statue of King Nebuchadnezzar II, and if they did not worship this golden statue, then they would be thrown into a fiery furnace.

This story stands out to me because here is three men who stood up against a king, who stood up against his tyrannical government and said, "This is not who we are; we're not going to worship you, for we worship the one true God. If it is His will for us to be thrown into the fiery furnace and perish, then so be it."

These three men stood up against the king. They did not kneel. The king was upset! These men were his advisers, even friends, and I think he took more offense because of that... So, what happens next? The king increases the temperature of the furnace seven times over. He says, "I'm really gonna turn the heat up on you guys, so you guys are really gonna feel how bad an idea it was to defy me."

The story goes that when the king's guards, the mightiest men in his army, were leading these three men into the furnace, they themselves were overcome by the heat and burned. Shadrach, Meshach, and Abednego entered the fire.

But they were not overcome.

They did not die. The king asks his watchman to look into the fiery furnace: what did he see?

He saw not three men there, but four:

"I see four men loose, walking in the midst of the fire, and they have no hurt; and the form of the fourth is like the Son of God." (Daniel 3:25, KJV)

God had sent an angel, perhaps Himself indeed, to protect these three men, His servants. Shadrach, Meshach, and Abednego were protected by God. They were then let out of the furnace and were promoted to power throughout the land.

Why should we all know this story?

There's going to be difficult moments in your life, moments in your life that are dark and that are going to try you. You may be put in front of a judgment, and you will be needing at that moment, *courage*. You will have to have the courage to stand up against what is wrong. If you are not ready, you'll never be able to survive the furnace.

See, the king is a lot like the mayor in my story. "Worship the vaccine. Do as I say. Doesn't matter if it harms anybody."

We know that this vaccination harms people. We're over three years out and we are seeing people lose their lives from sudden and unprecedented illnesses: diseases of the heart, strokes, seizures, weight gain, the list goes on and on. I never wanted to be right about it.

I just needed more data for myself, and I did not want people to be forced to take it. If you want to take the darn thing, take it! But don't *force* us to!

I was put into the fiery furnace, the hottest one. I was the only one that stood up. I was the only one that was terminated. I was the only one that saw these things and said something. I lost a career that I was passionate about. I lost that passion. I lost the opportunity to retire with those that I started with. I lost the opportunity to advance further in my career. All that was taken because of the decision I made.

For my decision to defend my firefighters, I was thrown into a fiery furnace. I survived because the Fire inside me burned brighter than the fires around me.

God is Walking with Us

My wife Veronika tells me that as people come into your life, they're either here for a *season* or they're here for a *reason.*

I will tell you that I have developed friendships all over the world, and those relationships were needed to strengthen what I was going through. Sometimes you're going to go through tribulations, and you need someone to be there with you. While I was the only person that stood up, while I was the only chief officer that said something, while I was alone in my decision, I wasn't *alone* because *God was with me the entire time.*

This book, *Alone in the Fire*, is about walking through tribulations. It's the story telling you that when you go through these dark times, these fires, you must know that God is with you. Have you asked Him to walk with you?

There is a picture that I have seen growing up. I've seen it in church. I've seen it at a private school. I've seen it in people's homes. I've seen it in people's offices. It is a picture of a beautiful beach. In the sand are footprints. The caption talks about a person who's questioning God, asking "God, why do the two sets of footprints here become one in the sand? This was my darkest time, the hardest time that I was going through. Why was I alone?"

God responds with, "There is only one set of footprints there because you weren't walking there at all, *I was carrying you.*"

I was not alone in my fire. I was with an angel who took care of me, who's allowed me to burn hotter and brighter internally than the fires around me, and that's how I've survived.

In addition to Him, there are all the relationships that I have. Without those individuals being a part of my life, during those moments when I needed help, I could not have done it alone. Many people supported me. Many people reached out. Those people were in my life for a reason.

Others were there just for seasons. Like my enemies, those who terminated me. I also lost friendships for my decision. Other relationships have ended because I've moved into different circles. My inner circle of people is different than those that I used to hang around with. Those people that stood with me – I can now stand with them in their own trials. They can always reach out to me, and I'm there as a person for guidance.

I've got more battles ahead of me. Greater battles. Wider territories. This book is going to ignite more movements. This book is going to start its own fires, ignite more people to be courageous in their communities. It's going to give them strength and boldness. It's going to put them in a position where they will say, "No, you're not going to do that to us. You're not going to oppress us. We will not kneel."

We will not kneel.

That all starts with one voice.

Defending our Nation

I'll end my chapter on this topic.

Individuals speaking up for what is right is foundational to our Western Culture. Consider how Christ spoke up for what is right, and how He was punished for that. Christ was ridiculed, spat on, crucified, killed. But not for long.

Consider how America was built. We the People stood up against tyranny. We stood up for what we believed to be right. We were put to the sword. We were oppressed. But not for long.

This value, foundational to our Western Culture, is under threat.

We're seeing this in our country. We're seeing this all over the world, and it's because too many people have been silent, have refused to stand up and to have courage, to have boldness. They're too afraid of what people may think or do.

During the pandemic, doctors lost their licenses. Lawyers lost their right to practice. Paramedics, Chief Officers, nurses, lost their jobs. Stand up against tyranny and tyranny will reply – in the only way it knows how.

But there is a power in *many*. If more people would be involved, if more people would push back and defend our values, we would not see oppression raining over us, ruling us in ways that shouldn't be. We have to believe that we're not alone in our fire, but that we are being carried *through* our fire.

We have to believe that we are being protected by God.

God gave me that internal Fire. God gave me the ability to stand up against the fires around me. God was my protective fire gear that I wear as a fireman in a hot fire. He was my helmet. He was my coat. He was my gloves and my boots and the air in my tank that I breathed. He was my mask that protected my face, the hood that protected the very hairs of my head.

I'll say it again, **we walk through fires when the Fire that is in us burns brighter than the fires around us.**

Chapter 6

The Raising Alphas Project

The Alphas: Stephen, Chris, and David.

**"Raising the Next Generation of Leaders not Followers" –
The Raising Alphas Project Mission Statement**

Remember back in chapter one how I spoke
about the red badge of courage?

This story was about me earning a badge for courage and bravery after seeing something and saying something. A little girl, about the age of three or four, fell into the pool. Her name was Alicia. I quickly responded by alerting her mother, Sue, who jumps into the pool, pulls Alicia up from the bottom, and embraces her. She comes over to me with tears and embraces me, saying, "You saved my daughter's life."

I was given a red badge of courage by Sue's husband, Tim, in front of my peers and leaders, as a Royal Ranger. I remember that this day was the reason that I went into public service. This is the reason I became a paramedic, why I joined the military, why I became a fireman, why I promoted so quickly, it was all so I could help others.

I always want to be the one that comes to the rescue.

So, what did I do when that ability to help others was suddenly and swiftly taken from me?

I quickly realized that there was an emergency all around me, one that I'd never seen before, one that was threatening the very foundation of our country and our culture: our children.

This is where The Raising Alphas Project begins.

The Launch

My business partner and friend, David Panzik and I, came up with the idea to launch a company that focused on raising strong kids. We knew that this was needed right now. He and I had been friends for years after meeting in the boxing ring at our gym. David had been in the fire service and left; I had been in the fire service and left. We realized very quickly that there was more to being a public servant than just working for the government. David and I had been going back and forth on getting into business. We had a mind shift which turned into a mindset. That's where we still live today, in a different mindset.

It was around 2020, right as the pandemic started to kick off, and the lockdowns were beginning. I have two young boys around the same age as David's two boys. In that pandemic, we saw our boys being attacked, and who was there to speak for them?

We couldn't go inside the school with them. We had to force face shields or masks on them. We were governed by people telling us that our children could kill them if they didn't have a mask on. Amid this, David and I wanted to serve something bigger, something greater. We knew that we had to stand up. We had to be vocal.

That's exactly what we did. We came up with the name, The Raising Alphas Project. It started out as a shirt idea, an eye catcher. I remember the day David and I first made it public. We attended a Save Our Generations conference in Orlando. On the way there, with this box of shirts in our car, we pulled into Lowe's and bought a little tool cart to put all the shirts in. We weren't prepared! We were so green, we weren't thinking ahead like, where do we put the shirts that we're going to sell?

We put the shirts on and started walking around the convention center. Dave and I are pretty big guys, and we stand out as we're walking, wearing our shirts that say Raising Alphas on the front. I remember we got so much attention that we got the opportunity to sell all our shirts at a booth that a vendor hadn't shown up to but had paid for. It was an open table, no one was sitting there when we showed up to it, so we'd settled in. We settled into the opportunity we saw.

There were lots of people in the convention that were selling their items, selling their books, their CD's, selling all-natural supplements. We got to join them and sit at that table for free. I checked repeatedly with the convention, "Are you guys sure this is ok?" They said, "We want you guys to be represented here and we want you to sell your shirts."

How is that possible?

Praise God, and that'll tell you!

By His grace, the graciousness of the convention managers, and the quick thinking of ourselves, two guys were able to show up with shirts and suddenly given a $600 table... We put our shirts out on that table. We were so unprepared: we went there with no cash; we had no form of payment; we couldn't even give out change. We're literally downloading the Venmo app on the way there!

We didn't have a cash box. We had a cardboard box!

That's how Raising Alphas started, with two men at a God-given booth who were sick and tired of listening to authorities tell us that we must mask our children. I wrote earlier about not being able to attend my oldest son's kindergarten graduation, or to walk him to class on his first day of school – something a parent dreams about. There are parents that didn't get to see their children graduate high school. There are high school kids that didn't have proms or homecomings or get to play in the last football game their senior year.

The Raising Alphas Project was about raising the next generation to be leaders, not followers, so that the 2020 tyrannies would never happen again.

Today, David and I have a podcast show, **The Raising Alphas Project**, that is seen not just in the United States but all around the world. We discuss important topics surrounding the challenges of raising the next generation of alphas, and we bring on guests to hear their stories of similar topics.

People resonate with our message. They resonate with the idea of being more involved in your kids' lives, teaching them early to be respectful, getting back to those family-first values. I can tell you that it's important to raise the next generation to be leaders, because if we don't have leaders in our future, then who will fight for us?

There's an analogy we use on our show, and the analogy goes that, one night, an intruder enters your family home – who responds to the emergency? As a father, my responsibility is to be the leader and the protector. If I have an intruder come into my home, he's going to meet force. He's going to meet a peaceful warrior suddenly throwing off that peace, revealing the fury underneath. Next in the line of defense is my wife. After that, my sons. And when I am too old to defend my home effectively, my sons become that first line of defense.

There's a greater lesson here. The home, in this analogy, is our country. My sons are the next generation.

Is the next generation ready?

We need leaders not only in the household, but in the community to stand up against evil, against tyranny, to stand up against the things that are wrong, so that when they see something, they say something. This is the importance of a leader.

Whether you're in the fire service or whether you're in law enforcement or whether you're in government, you need to be a leader. We need leaders in business, leaders for the everyday things that we go through, because the threats to us are everywhere.

Are you being a leader today?

Defending our Values

Our western values are under threat. They've been under threat for a while now. We can say that our western values have been under siege since 2020, but we could go back way further – this didn't just start happening. This has been going on for decades by governments and their corrupt leaders.

It's become more prevalent today. Those individuals that sit in office today do it without any thought, without any conscience, as if it's normal to violate people's rights. In 2020, they forced us to go home. They isolated us from our families. Our government isolated us from our loved ones, our friends, isolated us from our jobs, and from our churches. They took away things that suppress stress, like going to the gym, like being outside in the park enjoying the air, like going for a walk on the beach. How many parks were closed off from our kids? How many youth activities were canceled? The Orange County mayor shut down youth activities but left his cash cow convention center wide open.

Our western values are under threat because they forced medical procedures on us. Think about this, your loved one is in the hospital because they're sick, and who would they look to when a procedure comes? When you're approached by a position of authority like a doctor, and they say you must get this procedure, you must get this medication, and you're sick, you're not thinking right, so you look to your family to help you decide, but your family is not there – who do you look to?

Your family's the one that's supposed to take care of you. But during this time in 2020, they did not allow people to be with their loved ones when they were sick, and our sick loved ones were forced to make decisions under duress.

The corrupt have infiltrated political parties, law enforcement, the judicial system, and the education system. They've infiltrated medical facilities, our high positions of authority, to coordinate, target, and mandate drugs.

I can tell you that the upcoming 2024 elections are going to be the most important ever. If you're reading this book after the Presidential election in November, I hope that something has come out for the good, and that good has come from this book! Maybe people read this book and started to become more aware of the severity of our situation. I hope if you're reading this book after 2024 that you can say, "Yeah, we're in a good place now, things have gotten better."

We are in a moment of facing the unknown. We're in a moment where our western values are under threat and our own government is the one that's threatening us. They're putting things in place to confuse us. People are afraid to lose their jobs. People are afraid to make decisions. People are getting vaccinated not because they want to, or think that it's gonna help, but because they're scared. They don't want to lose their profession, their career, something they have worked so hard for.

But we lose something greater than our careers when we lay down our values for the illusion of safety. When we lay down our values, we also lay down our liberty, our power, and our dignity. Is all that worth a job?

Looking Ahead

What's next for The Raising Alpha's Project? What started as a local thing is now getting more and more traction, something that's being heard all over the U.S. and the world. We are international. So, what's next?

Books. I intend to write books, more books on this project, leadership books, books that will help young men and women be prepared for when they go into the real world, books on finances, how to be financially free, how to be an entrepreneur and a small business owner.

I want to break through the three taboos in America, the three things you don't talk about at the dinner table – *money, religion, and politics.* If we would start discussing those things, we'd get better at it, maybe not be so scared of it.

I've spoken about it earlier in this book, that the only way I got better as a fireman was, I took advanced classes, I challenged myself. Similarly, we need to advance our education in these areas. We don't ask the right questions about finance. We don't ask the right questions in politics. We don't ask the right questions about our religion because it's considered taboo.

I want to make sure that our young men and women have been properly educated on these topics. I want them to be involved with their parents' lives. I want those parents to be involved with their children's lives. I want us to get away from the technology that runs our world today. I want us to engage with the world with clarity, sagacity, and integrity.

That all starts with how we engage with our household. How we engage with our household sets the blueprint for our engagement with our communities, our country, and our world.

How to Make a Public Records Request

I want to end this chapter by giving you, the reader, information on one of the most valuable tools that we as individuals have against tyranny in our government: A public records request.

How do we make a public records request, and why does it matter?

Any type of government entity is required to keep public records, whether it's on the cell phone, whether it's on the computer, whether it's a paper document. It's all recorded, and what is recorded is accessible to us. Public records are something that we the people use as a form of checks and balances, and when people use these public records to check and balance our government officials, we keep them in line.

But what we've seen over the years is that people are becoming complacent. People don't even know what a request is or how to make one. This is exactly what corrupt elected officials are hoping for, that we don't hold them accountable. We don't check and balance the system because we refuse to be involved. We refuse to stay vigilant. But we can change that, and we can change that starting today.

A public records request is an easy form to fill out. You can go to your local government agency, whether it's a city, a town, the state, or the federal government. Go to their website. Find out where their public records request forms are, usually it's an e-mail. Once you find that e-mail, you can request information, things like:

- Looking up an email exchange between a government official and a business or other government employee
- Public Hearing and minutes. This can include video/audio recordings.
- Body Camera footage for officers that may have conducted themselves inappropriately.
- Official Reports or Government documents.

Now, in most cases, you're not required to give them any of your information. You can be completely anonymous. When you request for public records, ensure that you are not asking for months and months of information in one request, because this can be costly. Separate your request. Keep the request under 15 minutes of work for the employee conducting the search. This will save you money.

My lawsuit against Orange County began because of a public records request.

I'd lost my Step 3 grievance on the 28th of February, the day after my 40th birthday. I was so enraged, I began doing public records requests. I began to hold them accountable. I became the check and the balance.

I requested public records for all the disciplines that were issued because of vaccine noncompliance. Their response to me was, *there are none.* I laughed when I got that e-mail, because Danny Banks (Orange County Public Safety Director) had told me that the reason that he was denying my Step 3 grievance was because *there was not one individual who had been issued discipline improperly.* I emailed Danny these records and his response to me was, 'That can't be right, I'll get back to you.' Three hours later, he got back to me saying I had eleven reprimands in my inbox, but there were twelve people to be reprimanded. So where was the twelfth?

In addition, all eleven of the reprimands didn't have signatures from human resources, so they had never been properly accepted by human resources. They had not been properly filed in the government public records. Were they violating the law by not filling that out? This was six months after they were issued. I told this to Danny.

No response.

I began to dig deeper. I called everybody on the reprimand list, the list used to track all employees who did not comply with the mandate, over one hundred names, I called them all. I asked every single person to start doing public records on their own personnel files.

We found out that my friend, Dan, who had filed for a religious exemption form, had been issued a discipline. I was amazed. I had been told by my union leader, Paul Riccardi, that Dan had never turned the exemption request in. Dan was told that the written reprimand would be thrown out and not put into his personnel file. Dan told me this on the phone. He stated that it never went into his file. He never checked. He never followed up with it. So, when I did, he was confused on what I was asking him, because he was told he was good.

Dan showed me the e-mail response from human resources confirming that he had in fact filed for his exemption. I pulled another records request and found that Kim Buffkin never forwarded Dan's exemption to anybody. Nor did she reply to my replacement who had asked what to do about exemptions. She did not present any of this information to my Step 2 or Step 3 grievances. Doing so would have exonerated me. Doing so would have given me my job back.

She failed to be a leader. The evidence was hidden. This is the corruption that we live in. This corruption thrives because no one checks and balances these people.

I ended up doing more digging, and I found more things wrong with those disciplines. Orange County began to sign them about seven days *after* I inquired about them. Therefore, the disciplines were officially filed in human resources six months after the fact, *when it was already illegal to do so as decided by the Governor of Florida.* If you don't think that this is wrong, then you need to check yourself, because this behavior is flagrantly against how public records are kept in this country. *I filed my lawsuit because I found out that they had evidence that would have exonerated me, and they buried it.*

They were trying to bury me with it. They failed. This is just one piece of a very large pie. After reading this, you should turn to the back of this book, pg. 199, to the Appendix. Look at these emails. They're clear as day. Public records are how we check and balance the people who are given authority over us. As for me, I will hold my government accountable. Are you doing the same?

Chapter 7

Be All You Can Be

Sergeant Stephen Davis, Flight Medic & NCO

"Americans love a winner and will not tolerate a loser.
Americans play to win all the time. I wouldn't give a hoot in
Hell for a man who lost and laughed."
- General George S. Patton U.S. Army

In The Raising Alphas Project, our mission is to raise the next generation of leaders, not followers. How do we do that as parents?

At Raising Alphas, we teach our kids good family values, morals, and ethics. We teach them to be good people. That's what we want to start with, raising our children to be good people, able to recognize things that are wrong. When we're able to teach them what's wrong and right, what's good and evil, then they can have the understanding to ask questions and challenge authority.

At Raising Alphas, we raise up our children to be that authority, to be leaders.

Take for example an aspiring captain for a football team. What does an *alpha* captain look like? A captain's responsibility is not just to win the game. It's to bring out the best in others, to motivate the team when they're down, to press into a motivational intensity whether winning or losing.

Tim Tebow is a great example of this. After an infamous lost to Ole Miss in 2008, destroying the hopes of an undefeated season, he said this:

"I'm sorry. Extremely sorry. We were hoping for an undefeated season. That was my goal, something Florida's never done here. But I promise you one thing, a lot of good will come out of this. You have never seen any player play as hard as I will play the rest of this season, and you'll never see someone push the rest of the team as hard as I will push everybody the rest of this season. You'll never see a team play harder than we will the rest of the season. God bless." [2]

Later, when Tebow led the Florida Gators to their second victory as a student at the University of Florida, his level of intensity was still like no other. Where did this come from?

Well, ultimately (as he himself would say), his strength comes from God. But his home values were strong. His faith was rooted deep in the family. His parents raised him to be a leader. What a difference that made.

I believe that if we as parents engage in our children's lives each and every day, finding those simple things that they might be struggling with, and praising

[2] Hazarika, P. (2023, August 24). Remembering Tim Tebow's "the promise" speech which inspired Gators to the 2008 SEC championship. Sportskeeda. https://www.sportskeeda.com/college-football/news-remembering-tim-tebow-s-the-promise-speech-inspired-gators-2008-sec-championship

them for what they are doing well, we can help fulfill their fullest potential. To do this, *we, like our children, cannot be afraid to fail.*

Have you ever heard the phrase, *failure is not an option?*

Well, it is. Failing is an option. But when you fail, you learn. So, one way to look at failure is this, "I've never lost, I've only *learned.* So, our options then become – **I win, or I learn.**"

Failure is not an option, but learning from failure is. And in order to be a great winner you have to *learn* a lot.

First Lieutenant Godwin

In my life, I have served under some of the best leaders. I have also served under some of the worst leaders. I have learned from both.

When I went into the Army, I met one of the best. The army's motto was 'Be all that you can be.' It was only under the leadership of First Lieutenant Jeffrey Godwin that I learned exactly what this meant.

First Lieutenant Jeffrey Godwin, United States Army, was in 1st Division 2/2 Infantry. He was my mentor. He was a leader, and he was a friend. He moved up the ranks on the enlisted side of the army and went to school to be a physician's assistant.

When he was promoted to officer, his military career consisted of Ranger School, Airborne Special Forces, Green Beret, many tours and deployments to combat and peacekeeping missions, and other operations. Lieutenant Godwin was a well-respected soldier, officer, and clinician. He earned this respect from the person and leader he was every single day.

The team approach to everything was the most iconic thing about Lieutenant Godwin. He believed that everyone doing their part on the team can hurdle any obstacle. One person giving up on a difficult task only makes the weight even heavier for the team. Whereas everyone reaching in unison for the same goal makes that goal attainable.

Lieutenant Godwin taught us to soldier up by following the core values of the army, and I've mentioned this in a previous chapter, **LDRSHIP**. These values were the basic principles he instilled in us to better the platoon and ultimately the company.

BE ALL YOU CAN BE

A great leader creates great leaders.

Godwin created great leaders among the men under him by empowering them and affording them the tools they needed to be successful as mentors themselves. I can tell you that Lieutenant Godwin inspired me to be a better person, to be a better soldier, to be a better leader. As I mentioned earlier, I've had bad leaders, who I've learned from, but I've taken the most from my good leaders, because they gave me a pathway. They gave me direction. They gave me the opportunity to see things through their eyes.

I was given those things at an early age, 19-20 years old. Who was speaking into your life at 19? Who do you *wish* was speaking into your life at 19?

And whose life can you speak into today?

Lieutenant Godwin always pushed us farther. I remember I was working on my expert field medic badge. It's one of the most prestigious medical badges in the army. I remember finishing a 12-mile hike with a 50lbs ruck for my qualifications. As I'm heading towards the finish line, Lieutenant Godwin is running up from behind me and shouting, "Let's go, let's go, move it!"

He was pushing me farther.

Out of fifty medics in our platoon, in our battalion 2/2 Infantry, there were only two of of us that finished the run. To see the look in his eyes, the proudness in his demeanor when I crossed the finish line with one of my assistant squad leaders, was something I will never forget. He gave me a big hug, a big congratulations, and that to me exemplified the team approach.

I pushed myself farther for Lieutenant Godwin not because I had to, but because I wanted to.

That is my metric for good leadership. Are your followers working for you because they have to or because they want to?

I haven't seen Lieutenant Godwin since I left Germany in 2002. I haven't spoken to him. Still, his influence has been ingrained in me twenty-two years later. Because of him, I'm a better man. I'm a better person.

I try to stay true to what I've learned from him, and how I pass these values down to my children. Lieutenant Godwin was my role model. And I want to be a strong role model for my two boys. My boys are my legacy.

Who are the role models in your life?

Choose carefully because they are who you will become. We see young men and young women today, and their role models are people with little value at all. That's the importance of The Raising Alphas Project. We teach you how to be a strong role model for your children. That is how we make strong leaders of young men and women. We give them better role models.

For our children to be the best version of *themselves*, we must first transform us into the best version of *ourselves*.

A Leader is the Ultimate Team Player

I want to tell you one last story about Lieutenant Godwin.

One afternoon, outside during a thirty-day field exercise, I noticed that he wasn't eating. I asked him if he was hungry and if he was going to eat. He said, "I don't eat until every one of you eat."

He was over twice my age, over twice my authority, and yet he was waiting for us two-hundred men to eat before he got his meal. Godwin, and officers like him, were men that climbed so high in rank, men for whom we step out of the path, men that we salute as they walk by. And yet, when we're all coming down for chow, they are eating last.

Why are they doing this? **Because a leader is the ultimate team player.**

I would practice this too, this concept that a leader eats last. It was years later that Simon Sinek's book caught my attention, *Leaders Eat Last*. Simon is not a fireman; he's not a military guy; he's not a veteran, but yet, I find in his book this powerful concept and example of leadership.

Simon Sinek's *Leader's Eat Last* is one of the most exciting and inspirational books I've ever read. I've read it twice. I don't typically do that. Some books just find you in that way.

When I read this book, I found so many things that applied to my story, *Alone in the Fire*. I'm going to read to you an outtake from it right now, that talks about Germany in the 1930s, about living in Nazi Germany, and talking about following orders. This the common defense that many Nazis and Germans offered after the war, for 'just following orders':

We had no choice,' they said. We were just following orders, that was the mantra. Whether they were senior officials held accountable for their roles, or ordinary soldiers and civilians who tried to rebuild a sense of normalcy after the upheaval of the war, they were able to rationalize their actions, avoiding personal responsibility by holding their superiors accountable. This is what they would tell their grandchildren, we were just following orders.

During my last moments as a Battalion Chief in Orange County, so many other officers were just following orders. My superiors were just following orders, and this was their rationale.

The German Nazis were just following orders, but does that make it right?

No, it doesn't.

You should be held accountable if you're violating the law. You should be held accountable if you're violating people's civil rights. You should be held accountable, no matter the degree you were involved, whether it was issuing reprimands, or whether it was being silent.

Hannah Arendt, a German Jewish refugee and political theorist, once wrote, "The sad truth is that most evil is done by people who never make up their minds to be good or evil."

Ardent would write this while standing in witness to the trial of Adolf Eichmann, architect of The Final Solution, in Jerusalem.

We cannot afford complacency. We have to make up our minds today to be good. We must have the strength to be accountable for that good. Then we must have the strength to hold others accountable, no matter the social pressure.

If you were part of a government that was separating people from their families, you should be held accountable. If you saw something, and didn't say something, there's accountability for that too. When you allow things to just continue without addressing them, then it becomes normal, then it becomes something that is, 'Oh that's just how we do it.' I can't tell you how many times I've heard people say, 'That's just how we do it.' And it's not in policy, and it has no merit, but that's how we've always done it.

Simon Sinek also talks about how working in an unhealthy, unbalanced culture is like climbing Mount Everest. He writes about how we adapt to our surroundings, even in the most dangerous of conditions. Climbers will spend time at a base camp to adapt their bodies to the conditions of extreme altitude so they can persevere. We do the same thing in an unhealthy work culture.

Our adaptability is a two-edged sword. It can bring us to the summits of mountains. It can also acclimatize us to the lowest of hells. The conditions can be subtle, things like office politics, opportunism, occasional rounds of layoffs, and a general lack of trust among colleagues. We adapt. We get used to it. We let it be. Why do I bring this up?

Because in 2020 we were given slow changes, subtle changes, that eroded our freedom. The analogy I use is, you take a glass of water and you let it drip away one drip at a time and you leave it. If you were to sit there and watch it, let's say it's only one drip per day, you wouldn't really notice it. But if you were to come back six months later, you would notice the change. By then it's too late.

A leader has to recognize the leak before anyone else does, before it's too late. 2020 is when the leak started. Have we stopped it yet? Have we even noticed that it's happening?

Orange County's Hidden Evidence

We talked in the last chapter how the reason for my lawsuit was that my supervisor, Kim Buffkin, hid evidence for my Step 2 and my Step 3 grievances.

She had an e-mail from the Battalion Chief who replaced me which showed a Lieutenant had in fact a religious exemption and was acknowledged by human resources the day she relieved me of duty.

I remember it like it was yesterday. October 5th, 2021. It's about 10:00 in the evening, and everything that I explained to her that evening she put in an e-mail the next morning, and that e-mail was giving direction to all the other battalion chiefs saying, if you have someone who may be on the list who says they have a religious exemption, if you have someone who has a medical exemption, if you have someone who has already been vaccinated, let's confirm that this is correct *before we issue discipline.*

Therefore, everything I had explained to her about refusing to give discipline was confirmed.

She warned all the other battalions about it, yet I was suspended and then terminated. Later, on October the 19th, I found that the disciplines that other assistant and battalion chiefs had issued were returned to them, were torn up and were thrown away.

Now, if this is a government document, which is supposed to be a public record, why would it be thrown away?

Was it because of my digging? Was it because of my public records requests?

After I'd made my request, Kimberly Buffkin and several other assistant chiefs reached out to people that were given discipline improperly, apologizing for the mistake, promising removal of the reprimand from personnel files. I found these in public records.

Everything that I had brought to her attention and to the division chief's attention, that we should not be issuing discipline until it has been confirmed that these are warranted, was being covered up. They doubled down and tripled down. They hid this evidence from me.

When I discovered this, through public records requests, they corrected and protected themselves.

This is the power we have in holding people accountable. This is why it's important for us to be involved. I found these emails stating that everything that I had said was in fact true, but they did not want to present it. I was terminated for serious misconduct because I refused to issue discipline for people that did not deserve it. These emails confirmed that I was in the right.

The more public records I pulled, the more corruption I found.

Digging Down Deeper

I did a public records request and found that a Lieutenant, who happens to be the union president, Andre Perez, was arrested for assault and battery to his wife. What was his punishment?

36 hours of suspension.

He got a day and a half off. Why?

Another firefighter was caught on body camera striking a restrained, defenseless patient five times in the face. What was his punishment? He was removed from shift and placed into another position, as a mechanic, for five months.

Unlike me, both of these individuals remained employed by Orange County.

Why?

Well, I don't know. But the latter individual happened to be the son of a Circuit Court Judge. Coincidentally, this judge was on my case at the same time his son had committed this act of violence to a patient. I had to have this judge recused from my case.

But the corruption goes even deeper than this! When we really start looking into what has been happening around Orange County Fire Rescue, and many other municipalities during this time, around the country, around the state, around the world, we found more.

I will tell you all about what I discovered next in chapter eight.

Conclusion

I went into the military to be all I could be. I achieved that goal because I am still pushing myself to be better every day. I've made it a part of my life to be the best version of me, to make sure that I give my children the opportunities and resources to be better versions of themselves through leadership, through guidance, through coaching, through mentorship, through being a father. I'm raising the next generation of leaders in my household, and they will continue the legacy of the Davis name.

Chapter 8

Be All You Can Be, pt.2

Davis's Station 51 C-Shift. Picture for Mark & Todd Honor Challenge

"Befehl ist Befehl", aka the Nuremberg Defense
- International Military Tribunal, Nuremberg Principle IV

The Raising Alphas Project is coming into its second year in business, and we're growing faster every day. The need for us is there. This growth is fed by good people around us, good people that resonate with our message, like-minded people that know that strong and engaged parenting is what is needed right now.

We need to get this message far around the country. We need to grow this story. I can tell you that during these last couple years, and during the height of COVID-19, people have started waking up. This is how people started to resonate with The Raising Alphas Project.

With the incredible people we have on the show, I learn something new every time I host.

Guests on the Show

If we're gonna grow to be the best versions of ourselves, we can't be afraid to question the people in authority.

You know that saying, 'there's no dumb question.'? Well, people are afraid to ask any questions now because they are afraid to be ridiculed, afraid to lose their jobs and their credibility. What do we do about this?

My hope and purpose today is to inspire others to do something, to be an alpha. Each guest we bring onto the show is an opportunity towards that end.

I've had several great guests already on the podcast, such as my business partner, Chris Delgado. Chris is a leader in a different way. Chris is a young guy, younger than I by about 10 years. Yet he has taught me things as a grown man I never knew.

He's taught me things about finance investing. He's challenged me in ways that a leader should. He has opened a door for a different mindset. He's challenged me to read books that have opened a financial frontier.

As a soldier and a fireman, I have been successful because I put my mind to things. I work hard. I am driven, and I have a purpose, and I've achieved high positions during those times. As for entrepreneurship and becoming financially free to reach my purpose, to have more time with the family, to be able to provide with no stress, that was afforded to me because of Chris, a great friend, and a leader in my inner circle.

Another great guest, Austin Arthur. He's a business owner that started out in the fire service years and years ago, and now he's running for a commissioner seat. If he's hearing or reading this book after the election, I believe that he will be sitting in that commissioner's chair. If not, it doesn't change the fact of how he's brought so many people together in West Orange County through his business, through his leadership style. You can learn more about Austin's leadership style on our podcast.

Another guest, Ian Lord, a fellow firefighter at Orange County. He and his wife had a beautiful message to share about child adoption, but their story also involves me and Orange County!

Ian was out of country adopting their daughter in Nigeria. They went there in June of 2021. When they came back to the States in October, Ian was given a reprimand. (see p.214 appendix) Why was he given a reprimand? Because he didn't comply with the vaccine mandate!

Ian didn't get a shot, didn't turn in a card, didn't get a religious or medical exemption. How could he? He was in Africa the whole time! When Ian argued the reprimand with his Battalion Chief, that Chief told him, "I don't want to be Steve Davis'd..."

Steve Davis'd... I am a verb.

Another guest, David Iturrino. He and his wife, Annie, joined us early on in our show. He founded and directs a Jiu Jitsu gym. Something that he said caught my ear: *un-coachable kids become unemployable adults.* David has that hanging on a banner when you walk into his gym.

That's a powerful message. For Raising Alphas, how do we make young men and women leaders in today's society? They have to be coachable. Whether it's coachable on the field or in the classroom or at home, we have to start early on in explaining the why, giving direction, and instilling core values.

The coachable child becomes the coaching adult, the leader of tomorrow.

My Five Styles of Leadership

My personal leadership style stems from five distinct styles which I use depending on the situation. They are the following:

- The authoritarian leader, also called the autocratic.
- The delegative leader, also referred to as laissez-faire.

- The democratic leader, a participative leadership style.
- The transactional leader, also known as the managerial.
- The transformational leader, also known as the visionary.

I want to show you how and when I implement these styles:

When it comes time to decide as a Battalion Chief, working an emergency incident, I couldn't be the laissez faire type of leader. I had to be more autocratic and authoritative. I had to decide without input or giving power to anybody else. I had to decide and say, go to work. *Emergencies call for decisive leadership.*

Once this authority is established, and information illuminates the needs of the emergency scene, then I can start to delegate. I state the objective, and then I allow my junior leaders, my lieutenants, to make decisions. This only works if I have given them a clear objective. *This is the delegative leader.*

At the end of a fire incident, I always brought in all my officers, my lieutenants, and my captain, over to my truck. I would then implement the democratic leadership style. I would ask my officers, what am I missing? I would say, let's brainstorm. Even when a fire is still going, we can take a moment to discuss as equals. When everyone is oriented towards the same goal, everyone can contribute towards that goal. *That is the democratic style.*

The transactional leader, the managerial type, is something that I would set for long-term goals. On a day-to-day basis, I would set a target for employees to get certain things done. Whether it was training, cleaning their stations or their trucks or their uniforms, or to flow hydrants, or to do inspections, I set the objective and I allowed them to fulfill it on their own.

What's the transactional element there?

A reward or a penalty. If they do not meet the objective in a timely manner, there's a consequence. If they meet the objective in time, there's a reward. *That is the transactional leader.*

The transformational leader, the visionary. I can tell you that I've done this on a lot of different things. As a coach, I will get on to my kids about having a perfect pass. If they didn't pass the ball the way I'm asking for, I highlight it right there, but I also praise them and give them huge credit and huge enthusiasm when they get it right. I'm the coach on the sidelines that's jumping up and down, cheering, and high fiving other coaches and doing backflips. I'm the one that's getting intense on the side of the fields.

As Battalion Chief, I did the same thing. Well, maybe not doing back flips at the end of a fire... but I would praise people for the things that they've done well: Saving a life, getting a quick stop on a fire, finishing their hydrants within a couple weeks of the month, finishing all their training, completing their day-to-day objectives.

The visionary leader is the one who is ecstatic about the objective. The visionary leader is the one who feels most intensely the pain of failure and the sweetness of victory and mediates these feelings to his followers.

If you're only ever one of these styles, then you're missing the benefits of the others. What separates good leaders from bad is the ability to recognize which style is needed and when.

I've gone through good and bad leaders in a career that spans 2 decades. I can tell you that I've learned from every single one of them. How to be a great leader, how to be a bad leader, how I want to lead people, how I don't want to lead people, each was valuable to learn.

It all comes down to this, my personal leadership style is this, **get people to work for you because they want to, not because they have to.** At the end of the day, it's a better product.

The Power of the Positive

Positivity equals empowerment.

I believe in a positive environment. I believe that it is empowering when you can get to anything in your day and be positive. Of course, every single day there's gonna come up something that's gonna turn you down, but you have to look on the bright side. Does that mean walking around with flowers and sunshine all day long? No, but the goal is to remain positive even during the tough times. Why?

It's empowering.

I've spoken about it before, about positive and negative pressures. Too many times have I seen firemen at the fire stations come in with negative attitudes. Maybe they'll come into the fire station and complain about washing the truck. What!? You're washing a fire truck, and you're complaining about that? Or you're going on a call, going lights and sirens, and you're upset!?

Yes, I know you're tired, but let's look at the positive side of things. To wash a fire truck, to go on a call flying through red lights, lights and sirens blaring, helping people, saving lives, being a hero. With this positive outlook, the difficult aspects of the job become bearable.

Be positive when you have a room full of negative people. Be that beacon of hope. You're gonna get a lot of hate for it; you're gonna get a lot of laughs for it. Believe me, I have been that person. I have been that person that has been laughed at, been told I'm too positive. I can tell you that being positive will get you much further than being negative.

As I continue to climb the ladder of success in my personal life, I look down to see those individuals that have remained negative all these years, and I know that my positivity is what's allowed me to climb higher and higher than anyone else.

What do we do when we, as leaders, as bearers of positivity, encounter the overwhelmingly corrupt?

Battling the Bureaucracy

Have you ever heard the saying; absolute power corrupts absolutely?

As a positive person, I think that when people get into power, they for the most part have good intentions. I think that they want to be a part of something bigger than themselves. I think they want to be a part of something that's gonna make for positive change.

When I went from firefighter to Lieutenant, from Lieutenant to a Chief Officer, each time I promoted, I felt pride in what I was, and I wanted to do more for my organization. It was a process to get there.

I remember being a young fireman, on an auto accident, and I'm giving orders for the first time... *I felt strong.* I felt that this is what I was called to do. I remember that day.

It was a vehicle that had smashed into the back of a parked semi-trailer. It had messed up the driver pretty bad. We were cutting the vehicle apart, getting the passengers out, and I was giving directions on how to get them out safely and simply. Nothing too crazy, but it felt good, that power and authority and its responsibility. Power is responsibility.

Then, when you start getting into the higher positions, and you're not as checked and balanced, you feel the energy of 'I'm in charge', 'I'm giving orders.' It feels good, but there must come a balancing moment of humility. You must look in the mirror and say, I may be in charge, but I don't know everything.

I can't do everything as a fireman, as a paramedic. I can't carry the 300lbs man and save his life on my own. I need my team like my team needs me.

So, how do these leaders, who come into power with good intentions, become corrupted in these bureaucracies?

Temptation. It's that simple. People in power are approached by someone else that's been corrupted. Then that person is corrupted and corrupts another. It's like a never-ending loop.

How do they corrupt someone?

Maybe they got dirt on them. Maybe they promised them something. Maybe they load them up with finances, now they're wealthy, their kids are going to the best schools, their spouses don't have to work.

How is this environment of corruption allowed to continue? Well, it's because too many people sat around doing nothing. Too many Americans have sat around saying nothing. Too many Americans have never questioned the why, never gone against the grain, never challenged anybody, never held anybody accountable. They remained the *silent* majority.

Why? Because they just don't want to. They don't want to be involved. They just want to live alone and do their thing. How many times have you heard people say that? Maybe you're reading this book today, and your dream is to have 10-20 acres of property, and to be secluded away from corruption. Well, if you're not involved with making decisions in the community, then the community is going to be built by people who are easily influenced. When they're easily influenced, they can become corrupt. The cycle goes on.

Then it becomes very difficult to fight back.

How to be Immune from Corruption

When I talk to people who are running for public office, I'll ask them: Are you ready? Are you ready to take on corruption? Are you ready to be the only one to stand there and be alone in the fire? What's gonna keep you from being corrupted?

Can we inoculate our leaders, ourselves, from being corrupted? I believe it can be done.

Look at our President Donald Trump. Look at what they've been doing to him the last several years, even in just the last three years. He's not even in office again (yet)!

They do that to great leaders. The system is built to do that, but how do you stay strong? You build a strong community. You get rich. Part of The Raising Alphas Project is about learning new things, taking risks, not being caught up in the system. We do that with wealth.

Growing up, do we really learn about financial wealth and literacy? Do we really know how to do taxes. Do we really know how to open a business? Do we really know how to fund that business? We don't have any of this information when we're going through the public educational system.

Donald Trump is a good example of a leader that fights corruption because he's wealthy. He can maneuver around corrupting influences because he has wealth. He's built a strong foundation. If you're going to be a strong leader, money must be a resource and a tool.

Obviously, money can be used for evil things. It can be used to corrupt those into coming into power. It absolutely can. If I'm an individual with great ideas, if I'm an individual and a leader with good moral values, and I get into this system as a politician, how do I live virtuously when everybody is giving me money?

The solution is to come into power with wealth already secured. You want to bribe me with a Porsche? I already got one of those!

I know that sounds a little extreme, but that's exactly what happens. Politicians, the ones that come from wealth, the ones that have built their wealth, they're not easily corrupted. We can look at someone like Marjorie Taylor Greene. She's built wealth. Look at Tucker Carlson. Carlson can't be corrupted! He's got wealth, and he's even more powerful now that he's unchained.

That's how you create a platform. Create your wealth, your good foundation first, then take on power to do good. I think that's the best way. Build a foundation on goodness, virtue, and wealth. Stick to those values, and nothing can corrupt you.

Then you can be the change you want to see in your city, your state, and your country.

More of Orange County's Corruption

We talked about emails that had come out telling individuals that had received discipline that those disciplines were no longer in their file, emails that acknowledged the mistake. But let's go back for a moment to October 5th. On this day I was not only ordered to discipline people for being non-compliant with the vaccine mandate order, but I was also ordered to test each individual with a swab kit.

These kits were little boxes that almost looked like a Firehouse Subs to-go box. It was about that size and had a bunch of different items in it for testing. The boxes had already been opened and were taped shut. The expiration date on the boxes was crossed out, and a new expiration date was written on them in a sharpie.

I was ordered to have everybody tested with these kits. The problem is, I had multiple phone calls, text messages, and emails from lieutenants, firemen, engineers, captains, all saying that these kits were expired.

I brought this issue to Orange County leadership, and I was told to simply consult eighty-five pages of paperwork, fifty pages of which were lot numbers, to see if the kits were actually expired or not. This meant analyzing thousands and thousands of 7-digit numbers. Fifty pages worth.

This was but one of the many things that I was handling that day, including fielding phone calls mandating people to work. This is the tone that was set from the beginning of the pandemic: Confusion.

Keep us confused and we will not notice. We can give them medicine. We can give them written reprimands. They won't fight it because they're so confused and upset about the nasal swab. They're so upset about having to work under these conditions. Keep them confused.

I've been in this industry for over twenty-two years, and I can tell you that as a paramedic, you never issue or use equipment that is expired. That is something that has been ingrained in my very soul. Every other person in the department that I was fielding phone calls, texts, and emails from had the same belief. Each one didn't want to force their individuals to take these expired tests.

Let's be honest, there were officers that pencil whipped the tests through, that stated that tests were taken when they weren't. They were trying to protect their guys from using an expired medication and an expired piece of equipment.

How could I hold them accountable? I couldn't. I couldn't get a straight answer from my own leadership. I even called the company that manufactures the tests. After hours on hold, I only got bots on the line. And no documentation was given to me from them, nothing.

That was the setting that I walked into.

For those that do not comply, write them up, and let them grieve it later. Make them take it. Force them to take this shot. Force them to test.

This struggle is not over. Policies have been written in Orange County that will allow them to continue to do this. We're in the year 2024, and I can tell you that the policies that were put in place are still there. The governor's law, passed November 18th, 2021, expired June of 2023. What we know right now is that the lawsuits that were in play for Orange County have now been revived.

My fight is still on. The last bell has not rung. I intend to win it. I intend to be the last one standing when the last bell rings.

Chapter 9

The Five-Rung Ladder of Leadership

Davis with his Battalion 4 C-Shift Officers. LADDR!

"Leaders must own everything in their world. There is no one else to blame." - Jocko Willink, Naval Seal Officer (Ret.) in his book, *Extreme Ownership*

The five-rung ladder of leadership stems from my five styles of leadership in the previous chapter. But where the five styles were strategic tools – knowing which leadership styles to apply and when – the five rungs are more like a learning process a leader will go through. Each rung of this fire ladder is an ascent towards becoming a better leader:

Rung 1. Legitimate

Rung 2. Align

Rung 3. Dedicate

Rung 4. Develop

Rung 5. Rally

In this chapter I'm going to break down each rung, what it means to me, and how I've used it to become a better leader. This chapter is specifically dedicated to all I've learned so far on leadership, and now I want to share that knowledge with you.

Note that the acronym of the rungs spells out **LADDR** (ladder). That's easy to remember!

Rung 1. Legitimate

A leader, at this lowest rung of the ladder, is a title, a legitimate authority that people follow because they have to, not because they want to.

In the fire service, we need people to follow. We need people to lead. We need people to promote to positions of authority. The problem is many people in the fire service who promote become territorial.

They become influenced by their authority *as a position.*

Too many times have I seen "leaders" lead from the rear, and not from the front. They'll say, 'Do as I say not as I do', or 'respect the rank'. Too many times I've seen "leaders" touch their collar, slide their finger across their nameplate, referencing their rank. 'Remember who you're talking to.'

That's not real leadership.

There're too many people that have acquired their rank not because of their talent but because of quotas. It's not by what you've done or what you've accomplished or what you've put on the table to show value. No, it's because we must meet certain quotas of demographics.

How does that represent the County?

There're too many people in the fire service who have earned their positions not because they were the best person for the job, but because they were an individual who would just say *Yes.* They're yes people. They're an individual who will give direction based on an order that is improper, illegal, without blinking an eye, shedding a tear or a drop of sweat.

Rung 1, a legitimate leader's authority is based on the title, on the rank, not on their talent.

Rung 2: Align

At rung two, a leader aligns their followers to follow because they want to, not because they have to.

I learned this very important rung on the ladder of leadership in 2003 from a pilot in the military, a pilot by the name of Mike. I worked for him on a Black Hawk. Mike was the one that would tell me that phrase:

Get people to work for you because they want to, not because they have to.

I adopted that saying from him over twenty years ago. I've used it for twenty plus years. If people work for you because they want to, in the end you'll get a better product. I believe this deep down in my heart.

The Align rung is probably one of the most important. To even reach it at all, you must spend time with people. You must develop relationships. The reason people followed me was because they wanted to. I spent time with them. I built relationships.

People have asked me, "Who was your favorite crew?" There were crews that I spent more time with, yes, my station crews where I was a Battalion Chief. Those crews I was a little closer to, but that is because we ate dinner regularly together. That's where I laid my head down in the middle of the night. Those were the individuals I ran the most calls with. But as far as a favorite crew is concerned, I spent as much time as I could with everyone.

Everyone was equally important to me.

I set up breakfast, lunch, and dinner times. I would go and work out and train at every busy station. I would make it a point in my day to speak with every fireman at that station. Not just small talk, I would spend time getting to know them, who's their wife, who's their husband, who's their kids, what are their hobbies, what do they like to do, how was their week, is there anything that's going wrong. I would spend time with them.

This is one of the most important aspects of leadership: taking the time to build the relationship. You must possess a genuine love and care for other individuals, for those people that you're overseeing as a leader. When they work hard, and that product is better because you have spent time building those relationships, you get a better outcome.

As a coach on the soccer field, I tell my kids, my boys, my soccer players, I'll say, "What's more fun than fun?". And their response will be, "Winning!" They would say winning because everything that we're trying to do on the soccer field is to win. Win on the attack. Win on the defensive. Win as a team. Win as a player. Win as a coach.

You can't win unless you have a good Alignment with the people you lead. I've seen too many officers come in, clock in, clock out, and not spend that time with their crews. They've lost that relationship. That's why their product was never good. That's why they weren't respected. That's why they, in the end, would not have a truly loyal following.

Rung 3: Dedicate

At rung three, a leader is dedicated to the organization, and this dedicates the followers.

If you have earned your position, not because you're talented, but because you're in the right place at the right time, because you're the right demographic, because you know somebody who's on the board, then you haven't done anything for the organization. People see it. When you have done something for the organization, people see that too. People see when you have put the organization first. They see your enthusiasm as a leader is for the organization, not just for yourself.

What have I dedicated to the fire service at rung 3? Well, this book isn't to highlight everything that I've done in my career, but I will give you a few examples to illustrate how rung 3 of the ladder of leadership works.

In Orange County, as a medic, I chose to work at the busiest station. I was known for that. There are firemen who have worked there much longer than I, but I did a pretty decent stint at the busiest of houses. As an officer, I worked in the busiest battalion. I was recruited to be the next paramedic coordinator. Why was I chosen? Because I had pushed myself; I was on the EMS Team; I was traveling around the state of Florida competing in EMS competitions; I was involved in special operations.

I went on to run the paramedic program. We had only fifteen students in paramedic school, but that grew to over two hundred. I was responsible for all of them. I created a better testing process for them. I created scenario-based practical's with mannequin scenarios, and big-scene scenarios, where the paramedics were put into very difficult situations.

In one case, I would put my paramedics in all ballistic gear, with helmet and vest, and I would give them a radio and the radio would have the dispatcher say, "Listen we just got a call, there's a woman in the background screaming." That was the unknown medical. It was training for real scenarios, for going into a gunshot or stabbing scenario... Or so it seemed.

When they arrive on scene, the woman was screaming not because she had been shot, but because she was pregnant. She was delivering. My paramedics had to deliver the baby. They also realized that the mother was a drug addict, so that complicates things. Is this scenario farfetched?

Maybe.

Was this an unrealistic scene?

Unfortunately, not.

I've run on scenes like this. I've run on a scene where we didn't know what we were going into. I've delivered children in the back of the rescue. This was not an impossible scenario. Why did I work so hard on these trainings?

I enacted rung three leadership for my organization because I believed in it. I wore my Orange County Fire Rescue patch with pride. I was dedicated. I wanted to see the organization grow. The paramedic program has grown to what it is today because others followed my leadership. They improved the program one hundred times over. *My dedication led to their dedication.*

The same thing happened for other programs! For example, I started the rescue diving program at Orange County. I created a product and a vision, and I took the right people and put them into place so they could enhance the program. I'll give credit to Brandon Allen, who made the dive program what it is today. He saw my vision and he took off with it. He had a passion, and now he is that individual that's creating level three dedication leadership for Orange County Fire Rescue.

Rung 4: Develop

At rung four, a leader develops their followers into leaders.

A true leader is always helping others. As I developed as a Battalion Chief, as a Lieutenant, as a medic, I knew that I was going to move up in rank. That was my purpose, to do bigger and better things, to improve the people around me, to improve the organization. To do this, I was practicing the art of rung four on the ladder of leadership, which is *developing people to be the best version of themselves.*

I have said for many years that I don't know everything, but I put people together to know everything. For example, I'm not an electrician, but if I had a firefighter who was an electrician before he became a fireman, he would become my expert. He would be the go-to person at a fire who I can ask, "Are we able to secure this power in a different way?"

This minor example applies to many things. For example, I once had an officer many years older than me. We were on a large trash fire together once. It lasted for hours. It was a big junkyard on fire. This Lieutenant came up to me, and he had set up on the scene, and as I got the scene secured, and started developing a plan of action and attack on this immense fire. Our plan was working. Then my Assistant Chief came in and changed the plan, creating problems with our original setup. There was a conflict in leadership.

The problem was that this Assistant Chief had never had an incident like this. Neither had I. But this much older Lieutenant, lower in rank than me, had done this type of call three times before, and he had learned from those times. I asked him for his advice. He proposed an effective system based on his experience.

When I went to my supervisor with this proposal, I got a lot of pushback. I got told, "No, go back to where you're assigned." I was then reassigned to a position of less authority on the fire.

We fought that fire for hours. Still no change. I went back over to my supervisor, and I told him again what the plan had to be. I stood firm. I said, "This is how we're gonna do it, chief, and I'm gonna tell you this is how we're gonna do it because we've done this before, and I trust my officers."

That assistant chief looked at me, and he was a friend, and he said, "Alright, this is on you though. I'm putting this all on you."

We went to work.

I said, "Danny, let's knock this down."

We'd been out there for hours already fighting it, but I knew Danny was a good leader, and I trusted him. This was rung four, Develop. This was putting someone in a position where they could best succeed, for themselves and for the team. I knew Danny was the most valuable asset on that fire because he had the experience, the knowledge, and the training.

A leader develops others where there is a need. Surround yourself with the ones who are loyal and who truly care about your success. The rung 4 individual is loyal to their rung 4 leader and that's because of what has been done for them to help them develop.

Rung 5: Rally

Rung 5 on the ladder of leadership is the top of the top. It requires a lifetime of excellence, of mentorship, of holding others up. This rung of leadership requires, first and last, *a lifetime of humility and self-sacrifice.*

Isn't that ironic? To be the ultimate leader, one must become the ultimate servant.

Right now, as I finish this book, as I write chapter nine, I hope that, maybe one day, I'll reach rung 5. Maybe one day, people will see who I am and what I represent. I've started an organization called The Raising Alphas Project. That's what I represent. *Alone in the Fire* is another ideal of who I am and what I represent. One day, I hope, people will seek me out, those that I've spent years mentoring and molding will become loyal. Is it achievable?

100%

Have I achieved it?

Not yet.

Rung 5 is the top of the top. A leader must earn this position by ascending the other rungs, ascending through legitimacy, alignment, dedication, and development. A perfect leader must spend time building relationships, must show what they've done for the organization, and must empower and develop others.

My stance against the mandate was protecting the brand, protecting my people. I wasn't doing it for myself. I wasn't doing it to be unruly. The mandate was wrong. It violated people's rights. We were issuing discipline improperly. We were putting public records into files and destroying them. We weren't doing our due diligence to confirm a list was correct.

There's a great show from HBO, *Band of Brothers*, that dramatizes the history of Easy Company, 2nd Battalion, 506th Parachute Infantry Regiment of the 101st Airborne Division in WW2. In the series, you follow the person of Colonel Richard Winters. You see how, as he promotes rank through the series, he wants to make decisions on the front line. Once, during a pitched battle, he wanted to take his guys and run the line and go on the attack. He was reminded very quickly by his Lieutenant Colonel to get back. It wasn't his responsibility anymore to lead from the front line. Winters had to trust in the leaders, the NCO's especially, that he had developed.

Winters had to stand back from the front line because he was the rallying point, the idea that his men were fighting for. And his men rallied to him! Winters was a rung five leader.

This is a very important part of the leadership ladder, because if we don't rally around other leaders like us, our organizations will never properly grow.

As a Battalion Chief, I was not going to be the first person on the scene. I was not going to be the person on the nozzle fighting the fire. I had to rely on my lieutenants. I had to rely on the fact that they had developed their firemen, who in turn had to rely on their training in fire school, etc.

A team is a ladder of trust and reliance, and the higher you move up on that ladder, like a chain of command, the more you must give of yourself. To achieve rung 5 leadership, you must have mastered the previous four rungs. That's why the best leaders are the strongest servants.

The ladder of leadership is five rungs strong. Each rung is important and essential to the ascent. Remember the ladder:

L egitimate

A lign

D edicate

D evelop

R ally

Now start the climb.

More Corruption Revealed

Three days after my termination, a Battalion Chief writes an email Friday October 8th at 7:37 in the morning, to Kimberly Buffkin:

"Chief, the write ups I sent last shift, are you going to hold them since they received the vaccine? I am having everyone fill out the form again whether they did it or not. I'm taking care of emailing all the people on the list with the excel sheet I sent you. *I didn't want to issue those that had vaccine a write up and then deal with the confusion afterward.*" (emphasis mine)

So, this Battalion Chief sends an email to the Assistant Chief who relieved me of duty the shift prior to ask, "Are we holding off on this? I have people who have the vaccination who have shown me their card and they're on the list improperly, do you still want me to write them up? This is going to create confusion."

Kim Buffkin, on October 8th at 7:53am, responds to this Battalion Chief, "I spoke to Chief Wajda on the 7th of October about these writtens and if they were being held, he stated *the writtens were being held by HR.*" (emphasis mine)

My concerns have been addressed at this point. My concerns of, "Hey let's hold off on this. I'm not refusing to issue discipline; I'm refusing to issue improper discipline," were being addressed.

This email admits that the list is incorrect. I saw something. I said something. I was terminated for it.

I didn't know that this e-mail existed at the time. During my predetermination hearing on October 13th, I couldn't use this evidence because it was never provided. Kimberly Buffkin sat there and said nothing about this. She didn't become a leader. She could have stuck up for me. She didn't.

Instead, she actively advocated for my termination. She advocated, along with the assistant chief, Mike Howe, to remove me from my position. They knew about this e-mail. They knew that the list was incorrect. They knew that there were people on the list that turned in accommodations, people who turned in their vaccination cards.

They had been advised by me first, and they had been advised by other battalion chiefs. Why am I the only one that was disciplined and removed from my position?

Because I was public in my dissent.

Many of the other battalion chiefs did not issue discipline when they were supposed to. Many simply waited it out. How are they not held accountable while I am?

Disciplines were signed days after my termination. This was a confusing moment in 2021, and it was supposed to be confusing. They got rid of me because they knew that I was a threat, a threat to their narrative, a threat to what they were trying to accomplish. They got rid of me because they wanted people to stay in line.

They got rid of me because of who I was. They knew what I was able to produce as a leader. They knew my influence. They knew that I was building an army, people who followed me because of who I was and what I represented. They knew that I was the only officer, and Chief Officer, who was listed in a lawsuit of over seventy personnel against the county fighting the mandates, fighting for our civil rights, *fighting for freedom.*

Now, at Orange County Fire Rescue, I am a verb. Battalion Chief, Captains, Lieutenants, Engineers, Firefighters and Paramedics, they all *don't want to be*

Steve Davis'd. That's what they've told me. They don't want to go against the grain because of fear of *being Steve Davis'd.*

If being Steve Davis'd means being persecuted for doing what's right, then I'm proud of my verb. I'll wear it all day. I'll wear it like a red badge of courage.

Chapter 10

The First Alarm is Now!

Chief Stephen Davis answers the alarm!

**"When you are wronged repeatedly, the worst thing you
can do is continue taking it; fight back!"**
- President Donald J. Trump

Our leadership at Orange County Fire Rescue
failed the responsibilities of the Covid-19 pandemic. It

was a failure in the ideal of everything our leaders should be. The employees of Orange County saw firsthand what they did to firefighters that got out of line. They saw what they did to me. The brand of the department is now tarnished. People are leaving it left and right. People don't want to be a part of the organization. The prestige is lost.

How do we prevent the corruption that occurred at Orange County from happening in our Country, among our families, among our world?

I wrote this book to speak out. The first alarm is ringing right now! Will you answer it?

What We've Learned so Far

I spoke in **chapter one** about my first call to action as a first responder and as a leader: the rescue efforts of a little girl by the name of Alicia. I saw something. I said something. She was saved. How many more people can we save by speaking out? *If you see something, say something!* It's that simple.

My termination by Orange County was the closing of one door, but also the opening of another, to a bigger and greater opportunity. A closed door is not rejection, but direction. Accept the doors that God is closing in your life, and step through the doors that He is opening, even if you are afraid. Especially if you are afraid.

I was ripped away from a passionate career that I longed to be a part of, that I had put everything into. Something like this could happen to you too. Anything you do, whether it's in your company that you work for, whether it's something you started, anything can change in a moment. Are you ready for the change?

In **chapter two** I told you about *the readiness*, how I prepared myself in fire school, how I prepared myself before that in military training and in the United States Army. *Leaders engage with the challenges that followers leave unattempted.* That's because they're not afraid to fail. To a leader, failure is only learning.

If I was to fail as a chief officer on a large structure fire, people could die, firefighters could die. Readiness must be treated with the utmost respect, because in the moments of greatest significance, a mistake can cost you your life or the lives of those around you. Make the mistakes now, learn from them, so that when the fatal situation arises, you will not fail.

When my team and I were put to the fireproof test, we did not fail.

Are you ready for your fireproof test?

In **chapter three**, I mentioned how the corruption in 2020 progressively got worse and worse. Just two more weeks... It's just a mask... It's just an arrow on the floor... It's just six feet... Stay home from church, from the gym, take a drink... eat more fast food... etc.

One by one, the drops drained out the bucket of our liberties.

I wasn't about to allow this to happen to my sons, to my family. I was not about to sit back and say nothing. I was in a position of leadership, and therefore obligated to say something. I was obligated to challenge those above me who were wrong. If you're wrong, you're wrong, regardless of your rank.

I will remind you that in 2020, Orange County mayor, Jerry Demings, shut down playgrounds while keeping his cash cow convention center wide open. Suddenly, a playground was against the law. Suddenly, walking on the beach was against the law. Who was going to say something?

He who would speak up against tyranny must be prepared to face its fires.

I was alone in the fire in **chapter four**. The day of my termination, I reported in on March 18th, anticipating a suspension and a reasonable reprimand. I refused to give reprimands for what I believed to be an unlawful order: forced compliance with a Covid-19 vaccine mandate. I thought my leaders had my best interest at heart. I was wrong.

I was alone in the fire they'd set for me.

I sat patiently waiting as they brought me back to tell me that my career had ended. Everything I put into this, blood, sweat, tears, time away from my family, all lost. Burned away in an instant.

They greeted me with handshakes and confused smiles. Why was I the only one fired? Because we're afraid. We're afraid they'll suspend us, that they'll take our rank away, that they'll take our jobs away, and for this fear we acquire and allow corruption. I knew the day that I walked out of that building might have been my last time I put on a uniform in the fire service. *But I also knew it wasn't the last time they were going to see me fight a fire.*

This book is just the *first* alarm. This book is to tell my story, but its sequels are going to uncover every bit of corruption across the state, across the country, and across the world. This corruption exists also in Australia. In New York. In Los Angeles. The corrupt are everywhere, smiling, scheming, with warm handshakes.

Will you stand with us against the corrupt? Will you join the movement?

In **chapter 5**, I told you about Shadrack, Meshach, and Abednego. They were not alone in the fire. They had a fourth person with them, walking in the midst of the flames, like the Son of God. They were in the most dangerous of all fires, with no gear, no air tank, no mask, nothing to protect them from the elements, but still, *protected*, protected by God's grace.

Why am I here today? Why am I able to write this book, to support my family, why am I more successful today than I have ever been?

God's grace!

Through Him, *the Fire inside me burned brighter and hotter than all the fires around me.* This was never about me. This was never about looking better in front of anybody. This was about answering an alarm of emergency. A fire was burning, and no one was putting it out.

I heard the alarm. I was answering the questions. I was fielding the concerns of my people, of my firefighters, of my battalion and others, phone call after phone call, I was fielding the questions that arose during the pandemic, questions my County was not properly answering.

Questions like, "How does a person respond to the levels of stress that we are going through?" A first responder sees people die, sees people at their worst, sees traumatic and horrific injuries, horrific scenes. The pandemic and its response added to that stress.

"What do I do about my family if I don't take the shot?" "What do I do if I lose my job?" "What do I do, chief?"

I had no answer. I couldn't get any questions answered by my leadership. They forced me to stand against them. They tried to force me to reprimand them. Instead, I took a higher position than any earthly authority: The authority of what's right, which is God's.

I wish I had been more brash! That's what they needed when I stood up to them in that meeting. I needed to be a little bit more brash to get my point across. Maybe I needed to be disrespectful... I wasn't. I walked in there with humility, wearing the mask. I walked in there with grace, shaking their hands. I walked in there as a loyal servant to the department, to the organization, and they terminated me.

I started this lawsuit because I found more corruption. They hid the evidence. All those emails I showed you they hid. They hid it to bury me and to protect themselves.

There were emails that said: "You were issued a written reprimand for not complying with the vaccination verification. Once an updated list was published, *it was determined you should not have received that reprimand.*" This was an email that they sent out after my termination.

Why wasn't this brought forward in my grievance case?

I'll leave you to answer that.

Chapter six, The Raising Alphas Project. I talked about earning my red badge of courage when I was a young boy in the Royal Rangers. I talked about being in the public service for over twenty-plus years. I talked about how all that was preparation for me to start Raising Alphas. Our mission statement is *to raise the next generation of leaders not followers.*

I want to engage with parents. I want to engage with children, young men and young women, to tell them that they are worth more than what they're being told. I want them to know that there are better things out there than just following orders without question. I want them to be leaders, to be financially free. My mission is to ensure that our children are raised in a good home, are raised with a mother and a father with good moral values, family values, Christian values. I want our children to able to respectfully challenge the things that don't sound right. *People resonate with that message.*

This country needs the next generation to be strong. Our western values are under threat today, and if we don't act now, they will be gone one drop at a time.

It's time to stand and believe that we can do something today. You must be engaged! *The Raising Alphas Project provides a platform for you* to help you raise your children up right. Give our podcast a listen today, watch our show, join our following.

Raising Alphas is in its second year. We're moving very fast. I knew at the end of 2023 that 2024 was going to be big. As I write this book, three months into the new year, we're taking on new opportunities every single day. We're sailing into uncharted waters. I'm excited and nervous. I know what it's like to be a soldier. I know what it's like to be a fireman. Now, as an entrepreneur, as a business owner, I'm uncomfortable.

Guess what?

Being uncomfortable keeps me going, keeps me motivated. I'm constantly moving. I'm constantly looking to see how I can help others. I've made connections all over the state and there's more to come.

For the parents, for the small business owners, for the leaders, every single day, remind yourself that our youth is the next generation of leaders. Remember this, *un-coachable kids will become unemployable adults.* This is why the message of Alphas is so important.

How are you leading them?

In **chapter eight**, we talked about the different styles of leadership: The *autocratic*, the *laissez fair*, the *democratic*, the *transactional*, and lastly, the *transformational leader*. I've seen them all. I don't believe that one size fits all. I believe that, as a leader, our followers respond best in different ways. Our task as leaders is to know which style to apply and when.

As a coach on the soccer field, I cannot coach some players the same as others. I must adapt. The army told me, *Adapt and overcome.* Adapt and overcome... That's what I did in 2021 when I lost my job. That's what I did in 2023 when I walked away from the fire service. That's what I'm doing every day as an entrepreneur, as a father, as a husband. Adapt and overcome.

In **chapter nine** we talked about the Five-Rung Ladder of Leadership. Remember the acronym, LADDR:

Legitimate

Align

Dedicate

Develop

Rally

I strive to be at that top rung one day. I hope that people will see me as an inspiration, somebody with experience, skill, and knowledge. To reach the top, we must first ascend all four of the lower rungs. What rung are you on?

Have you even started the climb?

If not, start today. If you're still climbing, keep climbing. If you're at the top, whoever you are, help up those below you (you already are). If you've reached that top rung of leadership, you're probably someone that I've seen and looked up to. Maybe you've written a book already. Whoever you are, I thank you for you, because you've created a pathway for us.

This book, *Alone in the Fire*, is about standing up against corruption, bearing the burden of that standing up, and finally, this book is about courageous leadership.

This book will live long after I am gone. This book is a part of my legacy. What will your legacy be?

The Lawsuit & What's Next

Two years after my termination, I still don't know what God's plan is for my lawsuit. What I can tell you is that as I close this final chapter, I can hear "The Second Alarm" already beginning to ring. It will expose more truths about the corruption we live in.

2024, we are in one of the most important presidential elections ever. As I write this book, I don't know what's going to happen at the end of this year. I can only hope that our state, our country, is in a better place than it is today. I hope that this book has made it big, and as you're listening to this on audio or as you're reading this, you're seeing the success of *Alone in the Fire*, how many people it has touched, how many people have been held accountable, how many lives it's changed.

I hope this book inspires you to be involved. Whether that's being involved in your local communities, running for elections yourself, whether it's in your HOA, whether it's as a business owner, whether you're on City Council, or a part of the school board, *take up the responsibilities your authority calls for*. There are people that are taking positions in society today who are there to control and to disrupt. They're disrupting our American culture. We need to respond to this alarm. It's ringing right now.

By reading this book, you are now not alone in the fire. We build on community. We build on the Rock. We build on strength, not cowardice. If you're reading this, I want you to use your phone and follow this QR Code right now. I have a special message for you:

Your Call to Action!

Become a voice! Stand up and be a leader! If you aren't a leader, become one!

I've given you the tools, the Five-Rung Ladder of Leadership, and the five different styles of leadership. Take up these tools. We are powerful when we stand together!

If just ten other battalion chiefs had stood with me and refused to give in to tyranny, the County would have had to retract their discipline. So, as you read my final words, know that **your call to action is to get involved with the Raising Alphas movement, to become a leader, and to help us raise the next generation of leaders.**

Help us help our children to be strong. We still have a big fight ahead of us. We still have fires spreading. We can put those fires out, but we need your help. We can stand up against the corrupt, to hold them accountable, but we need your help!

Do not forget 2020. It was the year our freedom was taken from us. Take it back!

I'm standing up. If I'm alone again in the fires to come, so be it. God will protect me. I will continue confidently in this my hardest battle because I am one of God's toughest warriors.

Join the fight with us.

THE FIRST ALARM IS NOW!

This is an internal battle, in our country, and in all our hearts, between good and evil. Stand up for the good. I call you to this action today. As you close this book, get involved, be engaged, be all you can be! Be a leader. Be strong and courageous.

Be an Alpha!

<u>Appendix</u>*

* Note, all material herein has been gleaned from the public record and is thus subject to the scrutiny of the citizens.

Emails

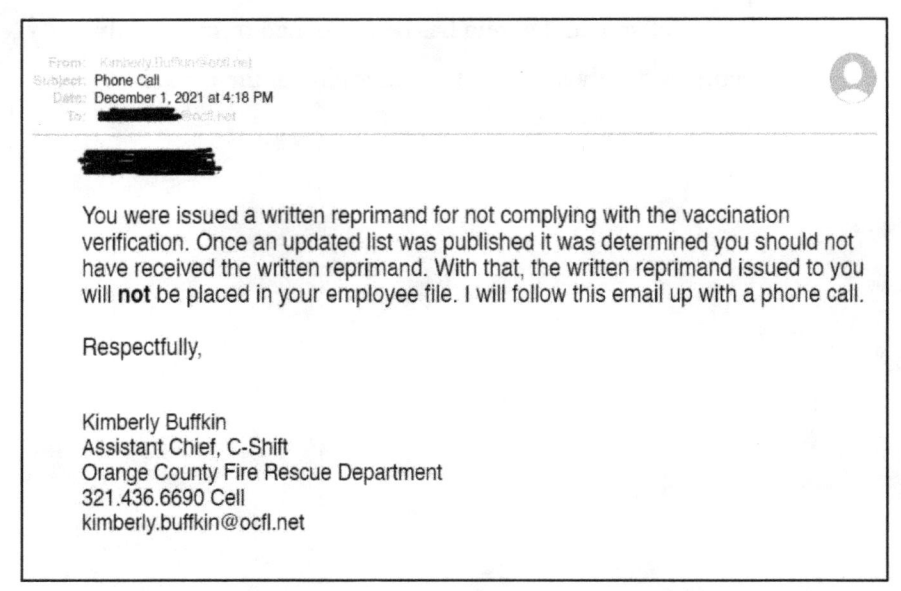

You were issued a written reprimand for not complying with the vaccination verification. Once an updated list was published it was determined you should not have received the written reprimand. With that, the written reprimand issued to you will **not** be placed in your employee file. I will follow this email up with a phone call.

Respectfully,

Kimberly Buffkin
Assistant Chief, C-Shift
Orange County Fire Rescue Department
321.436.6690 Cell
kimberly.buffkin@ocfl.net

A screenshot of an email from Assistant Chief Buffkin showing that reprimands were inordinately issued then removed from public files. The name of the recipient has been blocked out for their protection.

-------- Original Message --------
From: "Sherrill, Steven M" <Steven.Sherrill@ocfl.net>
Date: Wed, Oct 6, 2021, 3:00 PM
To: "Davis, Reginald C" <Reginald.Davis@ocfl.net>
Subject: RE: Accommodation

That's not what is happening. I have been issuing discipline today for those who claim they filed an exemption.

Thank you,

Steven M. Sherrill

Battalion Chief – B4 A-Shift
Orange County Fire Rescue Department
Office: 407-254-8451
Cell: 407-832-3381

From: Davis, Reginald C <Reginald.Davis@ocfl.net>
Sent: Wednesday, October 6, 2021 2:39 PM
To: Sherrill, Steven M <Steven.Sherrill@ocfl.net>; FR-Administration-Fire HR Service Center <FR-Administration-FireHRServiceCenter@ocfl.net>
Subject: Re: Accommodation

Sherril,

Prior to any discipline being issued, the department has been asked to first verify with the staff member if they have an exemption on file. So, you may be on the list, but after that verification is done, you would ultimately not receive any discipline if you submitted on or before Sept 30th.

Unfortunately, a lot of request came through on Sept 30th.

Thanks,

Reggie

-------- Original Message --------
From: "Sherrill, Steven M" <Steven.Sherrill@ocfl.net>
Date: Wed, Oct 6, 2021, 1:50 PM
To: FR-Administration-Fire HR Service Center <FR-Administration-FireHRServiceCenter@ocfl.net>,"Davis, Reginald C" <Reginald.Davis@ocfl.net>
Subject: Fwd: Accommodation

A screenshot of an email from Battalion Chief Sherrill stating that disciplines are being issued to those personnel who have filed exemptions.

<Katherine.Mora@ocfl.net>, "Gibson, Latarsha" <Latarsha.Gibson@ocfl.net>
Subject: Accommodation

Good afternoon,

I am on the list of employees receiving a written reprimand, however, I submitted my religious exemption on 9/30. Why am I receiving discipline when I sent my accommodation request by the deadline?

Regards,

Steven M. Sherrill

Battalion Chief – B4 A-Shift
Orange County Fire Rescue Department
Office: 407-254-8451
Cell: 407-832-3381

Email exchange: Orange County Battalion Chief Steven Sherrif receives a written reprimand.

10/08/21

To whom it may concern,

On 9/30 before 5pm I turned in my religious exemption to Orange County. On 10/05 I received a written reprimand the reason for the written reprimand stated "Per the attached COVID-19 directive, all employees were required to complete an on-line Covid-19 vaccination certification for by 8/31/2021 (extended to 9/30/21), unless otherwise applying for a medical or religious exemption. As of 10/1/21, County record reflect that you failed to complete the required on-line certification. All employees were reminded of this requirement on multiple occasions." Being that I turned in my religious exemption by 9/30 before 5 pm I should not have received this written reprimand. My signature on the written reprimand indicates that I have reviewed the discussion form but does not imply agreement with the action.

FF/EMT Vicente C. Cuevas #2572

77/C

An open letter to the County from Firefighter/EMT Vicente Cuevas raising his grievance regarding his written reprimand. Note that this grievance was raised before Chief Davis's termination.

RE: Religious exemption- Vicente Cuevas

Latarsha.Gibson@ocfl.net <Latarsha.Gibson@ocfl.net>
Sat 10/9/2021 12:36 AM

To: Vicente.Cuevas@ocfl.net <Vicente.Cuevas@ocfl.net>
Cc: **Covid-Mandate <Covid-Mandate@ocffa.com>**; Charles.Welch@ocfl.net <CharlesWelch@ocfl.net>;
Katherine.Mora@ocfl.net <Katherine.Mora@ocfl.net>

Hello/Good Morning,

We are currently working on ensuring that all certifications/exemptions have been received.

More information is forthcoming. My earnest apology for the delay.

Thank you,
Tarsha

From: Cuevas, Vicente C
Sent: Friday, October 8, 2021 9:32 AM
To: Welch, Charles R <Charles.Welch@ocfl.net>; Mora, Katherine <Katherine.Mora@ocfl.net>; Gibson, Latarsha <Latarsha.Gibson@ocfl.net>
Cc: 'Covid-mandate@ocffa.com' <Covid-mandate@ocffa.com>
Subject: Religious exemption- Vicente Cuevas

Good morning,
This is FF/EMT Vicente Cuevas #2572 I recently received a written reprimand for not being vaccinated. I turned in my religious exemption on 9/30 before 5 pm. Per IB 21-187 I was told to send an email to these email addresses if I believe I should not be on the list of employees to receive a written reprimand.

Thank you,
FF Cuevas #2572

PLEASE NOTE: Florida has a very broad public records law (F. S. 119).
All e-mails to and from County Officials are kept as a public record.
Your e-mail communications, including your e-mail address may be
disclosed to the public and media at any time.

An email from Firefighter Vicente Cuevas stating that he received a written reprimand regarding vaccine mandate incompliance despite acquiring a religious exemption. Note that this exchange took place before Chief Davis's final termination.

Lt Daniel Novoa
40/C

From: Joseph, Ronald <Ronald.Joseph@ocfl.net>
Sent: Tuesday, October 5, 2021 8:40 PM
To: Novoa, Daniel (Lieutenant) <Daniel.Novoa@ocfl.net>
Subject: FW: Accommodations Request Acknowledgment

From: Employee Relations <EmployeeRelations@ocfl.net>
Sent: Thursday, September 30, 2021 4:19 PM
To: Employee Relations <EmployeeRelations@ocfl.net>
Subject: Accommodations Request Acknowledgment

Hello,

This is to confirm we have received your request for an accommodation.

We are in the process of reviewing and assigning your requests to a Human Resources Representative.

Once assigned, the representative will reach out to you directly within the next week.

Attached is a document to assist with understanding **"What's Next?"** in the process.

Please do not respond directly to this email. The originating email account does not have any additional information on your request.

Thank you,

Employee Relations
Human Resources Administration
Internal Operations Centre I, 1st Floor
450 E South Street | Orlando | 32801

Email exchange confirming Ronald Joseph's religious exemption from the Covid-19 vaccine. Note appendix pg. 208 where he then receives a written reprimand, despite this exemption.

From: Buffkin, Kimberly L (Battalion Chief)
Subject: FW: F.F. Jeffrey R. Jarrell Weekly Covid Test Protest.
Date: October 20, 2021 at 4:03 PM
To: Wajda, Michael J Michael.Wajda@ocfl.net
Cc: Hepker, David A David.Hepker@ocfl.net

From: Phillips, Anthony D <Anthony.Phillips@ocfl.net>
Sent: Wednesday, October 20, 2021 4:02 PM
To: Buffkin, Kimberly L (Battalion Chief) <Kimberly.Buffkin@ocfl.net>
Subject: FW: F.F. Jeffrey R. Jarrell Weekly Covid Test Protest.

Good afternoon,
Please see the below.
Anthony D. Phillips Sr.
Battalion Chief
6th Battalion C Shift
Orange County Fire Rescue
Office 407-254-8467
Cell 407-627-8206
Anthony.Phillips@ocfl.net

From: Rodriguez, Juan <Juan.Rodriguez@ocfl.net>
Sent: Wednesday, October 20, 2021 11:46 AM
To: Phillips, Anthony D <Anthony.Phillips@ocfl.net>
Subject: Fwd: F.F. Jeffrey R. Jarrell Weekly Covid Test Protest.

Chief,
I'm forwarding you Firefighter Jarrell email per his request.
Respectfully,
Juan Rodriguez
Lieutenant 84/C

Begin forwarded message:

> **From:** "Jarrell, Jeffrey R" <Jeffrey.Jarrell@ocfl.net>
> **Date:** October 20, 2021 at 8:56:56 AM EDT
> **To:** "Rodriguez, Juan" <Juan.Rodriguez@ocfl.net>
> **Subject:** F.F. Jeffrey R. Jarrell Weekly Covid Test Protest.
>
> ?
> Lieutenant Juan Rodriguez.
> This email is to protest the Mandatory Weekly Covid Test I took this morning.
> I took the test to avoid progressive discipline. I request this email be sent to
> my chain of command. I have stated in past weekly covid tests protests that it
> makes no sense that only the Unvaccinated are tested when the C.D.C.
> shows that both Vaccinated and Unvaccinated can get and pass covid to
> others. I have yet to hear a good explanation for this and have to assume
> that it is to push an agenda and seperate those in this department as
> vaccinated and unvaccinated. I am deeply saddened that a highly respected
> member of this department was terminated this week for doing in my opinion
> what was the right thing to do. I serve Jesus Christ as my Lord and Savior
> now and for all eternity and he says in his word in John 16:33 "in this world
> you will have trouble but take heart I have overcome the world. In the first
> century A.D. Christians were fed to the Lions for standing up for their faith
> and in 2021 Christians who oppose taking a dangerous and untested vaccine
> are required to weekly test or face discipline and possible termination. The
> mandate from the President of the United States for the Vaccine is Unlawful
> and being pushed by those below him. I sincerely hope and pray this
> Department (and County) will live up to what some have called one of the
> best Fire Departments in the United States and let us have our freedoms as
> Americans that many Men and Women have died to provide for us.
> Thank you,
> Jeffrey R. Jarrell 1211, 84C
> 407-455-0800
> Jeffery.jarrell@ocfl.net

An email from firefighter Jeffrey Jarrell to Chief Kim Buffkin in protest of the mandate and in defense of Stephen Davis after his termination.

Just to confirm we are holding off on Lt. Coats' reprimand for now, correct?
I have his already signed in the event it needs to be issued.
JP

From: Buffkin, Kimberly L (Battalion Chief) <Kimberly.Buffkin@ocfl.net>
Sent: Wednesday, October 20, 2021 8:41 PM
To: Preston, Jared V <Jared.Preston@ocfl.net>
Subject: FW: New Verification and Refusal to Test Reprimands
Importance: High

From: Howe, Michael A <Michael.Howe@ocfl.net>
Sent: Wednesday, October 13, 2021 1:20 PM
To: FR-Assistant Chiefs <FR-AssistantChiefs@ocfl.net>
Cc: Wajda, Michael J <Michael.Wajda@ocfl.net>; Rathbun, David A <David.Rathbun@ocfl.net>
Subject: New Verification and Refusal to Test Reprimands
Importance: High

AC's,

Here are the new templates for the discipline related to the not completing the vaccination verification and refusing to test.

· **If there are personnel who have not received the written reprimand for not completing the vaccination verification ONLY use the one attached (you will have to hand write the names and employee ID in as the document had to be changed)**
· **For personnel who refuse to test use the applicable one attached**

If you have any questions, please contact me.

Thanks,
Mike

Michael A. Howe
Assistant Chief, Training
(O) 407-254-9005
(C) 407-948-0089

An email from Lieutenant Preston to Chief Kim Buffkin asking whether to hold flagrant reprimands. This demonstrates that Chief Buffkin knew Lieutenant Coats had a religious exemption but had been issued a reprimand regardless. This email was the date after Davis's termination. This evidence was not presented in his step 2/3 grievances. Had this evidence been presented, Stephen Davis could have been exonerated.

Emails between Chief Wendy Gross and Chief Kim Buffkin, showing that write ups were being held to double-check their validity. Note, this took place *after* Davis was suspended for raising that grievance.

From: <Wendy.Gross@ocfl.net>
Subject: FW: Lt. Patrick Connors
Date: October 20, 2021 at 9:26:19 AM EDT
To: <███████████████████>

Wendy J Gross
Battalion 5C/ 81
Office 407-254-8481
Cell 407-383-9506
wendy.gross@ocfl.net

From: Gross, Wendy J (Battalion Chief)
Sent: Friday, October 8, 2021 7:53 AM
To: Buffkin, Kimberly L (Battalion Chief) <Kimberly.Buffkin@ocfl.net>
Subject: RE: Lt. Patrick Connors

Gotcha… thank you I was verifying that as well. Enjoy your day off

Wendy J Gross
Battalion 5C/ 81
Office 407-254-8481
Cell 407-383-9506
wendy.gross@ocfl.net

From: Buffkin, Kimberly L (Battalion Chief) <Kimberly.Buffkin@ocfl.net>
Sent: Friday, October 8, 2021 7:53 AM
To: Gross, Wendy J (Battalion Chief) <Wendy.Gross@ocfl.net>
Subject: RE: Lt. Patrick Connors

When I spoke to Chief Wajda yesterday, he stated all the writtens were being held until HR could sift through the submissions of those that submitted late on 9/30. (This is the list I was talking about last shift that I was waiting for, the updated list). I am off today, I believe the direction will be to verify that the employee had not submitted either (vaccine attestation or accommodation request) by 10/1 and you are to have the employee sign it, but it will not be processed just yet. They have changed the plan because I believe they discovered several people submitted accommodation request the day before the deadline. This is changing daily, so I apologize for all the back and forth.

From: Gross, Wendy J (Battalion Chief) <Wendy.Gross@ocfl.net>
Sent: Friday, October 8, 2021 7:37 AM
To: Buffkin, Kimberly L (Battalion Chief) <Kimberly.Buffkin@ocfl.net>
Subject: RE: Lt. Patrick Connors

Chief,

The write ups I sent last shift, are you going to hold them since they received the vaccine? I am having everyone fill out the form again whether they did it prior or not. I am taking care of emailing all the people on the list with the excel sheet I sent you. I didn't want to issue those that had the vaccine a write up and then deal with the confusion after

Wendy J Gross
Battalion 5C/ 81
Office 407-254-8481
Cell 407-383-9506
wendy.gross@ocfl.net

From: Buffkin, Kimberly L (Battalion Chief) <Kimberly.Buffkin@ocfl.net>
Sent: Friday, October 8, 2021 7:24 AM
To: Gross, Wendy J (Battalion Chief) <Wendy.Gross@ocfl.net>
Subject: FW: Lt. Patrick Connors

Since he is going to be out for 3 shifts please make sure the

 **ORANGE COUNTY GOVERNMENT** FLORIDA

DOCUMENTATION OF
NON-DISCIPLINARY COUNSELING /
DISCIPLINARY ACTIONS

Employee Name: Ronald Joseph   EEID #: 132802

Department: Fire Rescue   Division: All

CURRENT NON-DISCIPLINARY COUNSELING / DISCIPLINARY ACTION: Date of incident: 10/1/2021

Non-Disciplinary Counseling* ◯   Verbal/Oral Warning (documented) ◯   Written Reprimand ⦿
*excluding LIUNA:attendance matrix

POLICY / CONTRACT VIOLATED:

Administrative Reg. #	Policy Manual # OCPM 401; 411
Operational Reg. #	Dept. Op. Procedure If applicable
Admin. Order# If applicable	Divisional Guideline If applicable
Standard Op. Procedure If applicable	Contract – Article # If applicable

SITUATION: See pg. 2 for additional space   Continued on page 2? Yes ☐ No ✔

Per the attached COVID-19 directive, all employees were required to complete an on-line Covid-19 Vaccination Certification Form by 8/31/21 (extended to 9/30/21), unless otherwise applying for a medical or religious exemption. As of 10/1/2021, County records reflect that you failed to complete the required on-line certification. All employees were reminded of this requirement on multiple occasions.

SUMMARY OF PRIOR DOCUMENTATION

Date: N/A	Counseling ☐	Oral ☐	Written ☐
Date:	Counseling ☐	Oral ☐	Written ☐
Date:	Counseling ☐	Oral ☐	Written ☐

ACTION PLAN TO RESOLVE PROBLEM	PERSON RESPONSIBLE	BY WHEN?
Wear approved face coverings in all work settings.	Employee	Immediately
Provide Proof of negative test 1 time each week.	Employee	Every Monday

Follow-Up Date: _____ Time: _____ Place: _____

CONSEQUENCES OF INSUFFICIENT IMPROVEMENT OR RECURRENCE MAY RESULT IN FUTURE DISCIPLINARY ACTION UP TO AND INCLUDING TERMINATION.

SIGNATURES:

[signature] 10-6-21			
EMPLOYEE* DATE		SUPERVISOR PRINT NAME	EEID #
HR REPRESENTATIVE DATE		SUPERVISOR'S SIGNATURE	DATE

* EMPLOYEE'S SIGNATURE INDICATES THAT EMPLOYEE HAS REVIEWED THIS DISCUSSION FORM, BUT DOES NOT NECESSARILY IMPLY AGREEMENT WITH THE ACTION. **EMPLOYEE HAS THE RIGHT TO APPEAL THIS DISCIPLINARY ACTION IN ACCORDANCE WITH POLICY AND OR APPLICABLE COLLECTIVE BARGAINING AGREEMENT.

Form 51-89 Revised 11/2018

Firefighter Ronald Joseph's disciplinary form. Note on pg. 204 his confirmation of receiving a religious exemption. Also note the section under 'Follow-Up Date', which is whited out in subsequent disciplinary forms after Chief Stephen Davis revealed that this policy contradicted #5 of the MOU, page 220.

DOCUMENTATION OF
NON-DISCIPLINARY COUNSELING /
DISCIPLINARY ACTIONS

Employee Name: Vicente Cuevas   EEID #: 132136

Department: Fire Rescue   Division: All

CURRENT NON-DISCIPLINARY COUNSELING / DISCIPLINARY ACTION: Date of incident: 10/1/2021

Non-Disciplinary Counseling* ○   Verbal/Oral Warning (documented) ○   Written Reprimand ⦿
*excluding LIUNA:attendance matrix

POLICY / CONTRACT VIOLATED:

Administrative Reg. #	Policy Manual #	OCPM 401; 411
Operational Reg. #	Dept. Op. Procedure	If applicable
Admin. Order# If applicable	Divisional Guideline	If applicable
Standard Op. Procedure If applicable	Contract – Article #	If applicable

SITUATION: See pg. 2 for additional space   Continued on page 2? Yes ☐ No ☑
Per the attached COVID-19 directive, all employees were required to complete an on-line Covid-19 Vaccination Certification Form by 8/31/21 (extended to 9/30/21), unless otherwise applying for a medical or religious exemption. As of 10/1/2021, County records reflect that you failed to complete the required on-line certification. All employees were reminded of this requirement on multiple occasions.

SUMMARY OF PRIOR DOCUMENTATION

Date: N/A	Counseling ☐	Oral ☐	Written ☐
Date:	Counseling ☐	Oral ☐	Written ☐
Date:	Counseling ☐	Oral ☐	Written ☐

ACTION PLAN TO RESOLVE PROBLEM	PERSON RESPONSIBLE	BY WHEN?
Wear approved face coverings in all work settings.	Employee	Immediately
Provide Proof of negative test 1 time each week.	Employee	Every Monday

Follow-Up Date: Time: Place:

CONSEQUENCES OF INSUFFICIENT IMPROVEMENT OR RECURRENCE MAY RESULT IN FUTURE DISCIPLINARY ACTION UP TO AND INCLUDING TERMINATION.

SIGNATURES:

(signature)	10/5/2021	_Emily A Elliott_ 112351
EMPLOYEE*	DATE	SUPERVISOR PRINT NAME / EEID #
		Emily A. Elliott 10/05/2021
HR REPRESENTATIVE	DATE	SUPERVISOR'S SIGNATURE / DATE

* EMPLOYEE'S SIGNATURE INDICATES THAT EMPLOYEE HAS REVIEWED THIS DISCUSSION FORM, BUT DOES NOT NECESSARILY IMPLY AGREEMENT WITH THE ACTION. **EMPLOYEE HAS THE RIGHT TO APPEAL THIS DISCIPLINARY ACTION IN ACCORDANCE WITH POLICY AND OR APPLICABLE COLLECTIVE BARGAINING AGREEMENT.

Form 51-69 Revised 11/2018

R006

Firefighter Vincent Cueva's disciplinary form. Note on pg. 203 his confirmation of receiving a religious exemption.

**DOCUMENTATION OF
NON-DISCIPLINARY COUNSELING /
DISCIPLINARY ACTIONS**

Employee Name: *Ethan Brodrecht*   EEID #: 132564

Department: Fire Rescue   Division: All

CURRENT NON-DISCIPLINARY COUNSELING / DISCIPLINARY ACTION: Date of incident: 10/1/2021

Non-Disciplinary Counseling* ○   Verbal/Oral Warning (documented) ○   Written Reprimand ◉
*excluding LIUNA:attendance matrix

POLICY / CONTRACT VIOLATED:

Administrative Reg. #		Policy Manual #	OCPM 401; 411
Operational Reg. #		Dept. Op. Procedure	If applicable
Admin. Order#	If applicable	Divisional Guideline	If applicable
Standard Op. Procedure	If applicable	Contract – Article #	If applicable

SITUATION: See pg. 2 for additional space   Continued on page 2? Yes ☐ No ☑
Per the attached COVID-19 directive, all employees were required to complete an on-line Covid-19 Vaccination Certification Form by 8/31/21 (extended to 9/30/21), unless otherwise applying for a medical or religious exemption. As of 10/1/2021, County records reflect that you failed to complete the required on-line certification. All employees were reminded of this requirement on multiple occasions.

SUMMARY OF PRIOR DOCUMENTATION

Date: N/A	Counseling ☐	Oral ☐	Written ☐
Date:	Counseling ☐	Oral ☐	Written ☐
Date:	Counseling ☐	Oral ☐	Written ☐

ACTION PLAN TO RESOLVE PROBLEM	PERSON RESPONSIBLE	BY WHEN?
Wear approved face coverings in all work settings.	Employee	Immediately
Provide Proof of negative test 1 time each week.	Employee	Every Monday

Follow-Up Date: _____   Time: _____   Place: _____

SIGNATURES:

[signature] 2615 10/14/21	Steven M. Sherrill	111732
EMPLOYEE* DATE	SUPERVISOR PRINT NAME	EEID #
	Steven M. Sherrill	10/14/2021
HR REPRESENTATIVE DATE	SUPERVISOR'S SIGNATURE	DATE

* EMPLOYEE'S SIGNATURE INDICATES THAT EMPLOYEE HAS REVIEWED THIS DISCUSSION FORM, BUT DOES NOT NECESSARILY IMPLY AGREEMENT WITH THE ACTION. **EMPLOYEE HAS THE RIGHT TO APPEAL THIS DISCIPLINARY ACTION IN ACCORDANCE WITH POLICY AND OR APPLICABLE COLLECTIVE BARGAINING AGREEMENT.

Form 51-68 Revised 11/2018

D 13 VIOLATION

A disciplinary form that was issued to firefighter Ethan Brodrecht. Note the area below Follow-Up Date has now been whited out.

A disciplinary form, issued to lieutenant Daniel Coats. Note the area below Follow-Up Date has been totally blacked out. Daniel Coats was in Davis's battalion and had filed for a religious exemption. See page 206 for relevant email regarding his reprimand, showing Buffkin's knowledge of Coats's exemption.

A disciplinary form, issued to lieutenant James Hansen. Note that here again the section under follow-up date, which stated that further noncompliance could lead to termination, is blocked out.

**DOCUMENTATION OF
NON-DISCIPLINARY COUNSELING /
DISCIPLINARY ACTIONS**

Employee Name: Brandon Ickes

EEID #: 129134

Department: Fire Rescue

Division: All

CURRENT NON-DISCIPLINARY COUNSELING / DISCIPLINARY ACTION: Date of Incident: 10/1/2021

Non-Disciplinary Counseling* ○ Verbal/Oral Warning (documented) ○ Written Reprimand ◉
*excluding LIUNA: attendance
matrix

POLICY / CONTRACT VIOLATED:

Administrative Reg. #	Policy Manual #	OCPM 401; 411
Operational Reg. #	Dept. Op. Procedure	If applicable
Admin. Order # If applicable	Divisional Guideline	If applicable
Standard Op. Procedure If applicable	Contract – Article #	If applicable

SITUATION: See pg. 2 for additional space   Continued on page 2? Yes ☐ No ☑
Per the attached COVID-19 directive, all employees were required to complete an on-line
Covid-19 Vaccination Certification Form by 8/31/21 (extended to 9/30/21), unless otherwise
applying for a medical or religious exemption. As of 10/1/2021, County records reflect that you
failed to complete the required on-line certification. All employees were reminded of this
requirement on multiple occasions.

SUMMARY OF PRIOR DOCUMENTATION

Date: N/A	Counseling ☐	Oral ☐	Written ☐
Date:	Counseling ☐	Oral ☐	Written ☐
Date:	Counseling ☐	Oral ☐	Written ☐

ACTION PLAN TO RESOLVE PROBLEM	PERSON RESPONSIBLE	BY WHEN?
Wear approved face coverings in all work settings.	Employee	Immediately
Provide Proof of negative test 1 time each week.	Employee	Every Monday

Follow-Up Date: Time: Place:

SIGNATURES:

10-8-21
EMPLOYEE*   DATE

Bertilus Bornelus 13554
SUPERVISOR PRINT NAME   EEID #

1228 10-8-21
SUPERVISOR'S SIGNATURE   DATE

HR REPRESENTATIVE   DATE

* EMPLOYEE'S SIGNATURE INDICATES THAT EMPLOYEE HAS REVIEWED THIS DISCUSSION FORM, BUT DOES NOT
NECESSARILY IMPLY AGREEMENT WITH THE ACTION. **EMPLOYEE HAS THE RIGHT TO APPEAL THIS DISCIPLINARY ACTION IN
ACCORDANCE WITH POLICY AND OR APPLICABLE COLLECTIVE BARGAINING AGREEMENT.

Form 61-69 Revised 11/2018

D.13 Violation

A disciplinary form that was issued to firefighter Brandon Ickes, also a member of Davis's Battalion.

ORANGE COUNTY GOVERNMENT FLORIDA

**DOCUMENTATION OF
NON-DISCIPLINARY COUNSELING /
DISCIPLINARY ACTIONS**

Employee Name: Ian Lord

EEID #: 112352

Department: Fire Rescue

Division: All   Operations

CURRENT NON-DISCIPLINARY COUNSELING / DISCIPLINARY ACTION: Date of incident: 10/1/2021

Non-Disciplinary Counseling* ◯ Verbal/Oral Warning (documented) ◯ Written Reprimand ◉
*excluding LIUNA: attendance matrix

POLICY / CONTRACT VIOLATED:

Administrative Reg. #	Policy Manual # OCPM 401; 411
Operational Reg. #	Dept. Op. Procedure If applicable
Admin. Order # If applicable	Divisional Guideline If applicable
Standard Op. Procedure If applicable	Contract – Article # If applicable

SITUATION: See pg. 2 for additional space   Continued on page 2? Yes ☐ No ☑
Per the attached COVID-19 directive, all employees were required to complete an on-line Covid-19 Vaccination Certification Form by 8/31/21 (extended to 9/30/21), unless otherwise applying for a medical or religious exemption. As of 10/1/2021, County records reflect that you failed to complete the required on-line certification. All employees were reminded of this requirement on multiple occasions.

SUMMARY OF PRIOR DOCUMENTATION

Date: N/A	Counseling ☐	Oral ☐	Written ☐
Date:	Counseling ☐	Oral ☐	Written ☐
Date:	Counseling ☐	Oral ☐	Written ☐

ACTION PLAN TO RESOLVE PROBLEM	PERSON RESPONSIBLE	BY WHEN?
Wear approved face coverings in all work settings.	Employee	Immediately
Provide Proof of negative test 1 time each week.	Employee	Every Monday

Follow-Up Date: _____ Time: _____ Place: _____

SIGNATURES:

EMPLOYEE*	DATE 10/12/21	SUPERVISOR PRINT NAME Brian J Beechner	EEID # 112815
HR REPRESENTATIVE Katherine Mora	DATE 3/18/22	SUPERVISOR'S SIGNATURE Brian J Beechner 1134	DATE 10-12-2021

* EMPLOYEE'S SIGNATURE INDICATES THAT EMPLOYEE HAS REVIEWED THIS DISCUSSION FORM, BUT DOES NOT NECESSARILY IMPLY AGREEMENT WITH THE ACTION. **EMPLOYEE HAS THE RIGHT TO APPEAL THIS DISCIPLINARY ACTION IN ACCORDANCE WITH POLICY AND OR APPLICABLE COLLECTIVE BARGAINING AGREEMENT.

Form 51-69 Revised 11/2018

D. 13 Violation

A disciplinary form that was issued to firefighter Ian Lord, who was adopting his children in Africa at the time the mayor's order was issued.

Termination Documents

Chief Davis's initial termination papers.

ORANGE COUNTY FIRE RESCUE DEPARTMENT
Michael A. Howe, Assistant Chief, Training
P.O. Box 5879
Winter Park, Florida 32793-5879
(407) 836-9000 Fax (407) 836-9138
e-mail: Michael.Howe@ocfl.net

Received:		Date:
Delivered:		Date: 10/18/21

October 18, 2021

Stephen Davis, Battalion Chief
6590 Amory Court
Winter Park FL 32792

Predetermination Hearing Decision Letter

Chief Davis,

A Predetermination Hearing was held on October 13, 2021, pertaining to you refusing to follow a direct order on October 5, 2021. During the PDH the following personnel were present: Assistant Chief Howe- Hearing Officer, Assistant Chief Buffkin- Observer, Assistant Chief Mack- Observer, Battalion Chief Stone- Observer, Battalion Chief Teamer- Observer, Charles Welch- Human Resources, Paul Riccardi- Local 2057, Steve Sherrill- Local 2057, Chris Newsome- Local 2057 and you.

I have carefully reviewed all written documentation; audio recordings as well as oral testimonies presented during the hearing and have made the following conclusions and determinations:

- You admitted you read GO 21-012 (Vaccine Mandate and Certification) that was published on August 20, 2021 and completed the Target Solutions associated with the GO

- You admitted you read GO 21-013 (Vaccine Mandate and Certification Update) that was published on August 31, 2021 but did not recall if you completed the Target Solutions associated with it

- You admitted you read GO 21-015 (On-site COVID Testing Target Solutions Training) that was published on September 30, 2021 but did not recall if you completed the Target Solutions associated with it

- You admitted you knew beginning October 4, 2021 the County was going to begin testing personnel who did not complete the vaccination verification

- You admitted you knew the County was going to begin issuing written reprimands to personnel who did not complete the vaccination verification form or file for an accommodation but you did not know they would begin issuing them immediately

- You admitted your first working shift after this took place was Tuesday, October 5, 2021

- You admitted Chief Buffkin discussed with you that personnel will be administering tests and reprimands would be issued to the applicable C-Shift personnel on October 5, 2021

- You were asked if you issued written reprimands to the personnel in Battalion 4 who were on the list to which you replied no you did not

- You admitted Chief Buffkin had a phone conversation with you on October 5, 2021 where she advised you it was a general order to issue the reprimands

- You admitted Chief Buffkin had you report to station 30 that evening with a Union Representative to further discuss the issue

- You admitted while at station 30, Chief Buffkin told you it was a direct order to issue the reprimands

- You admitted while at station 30, Chief Buffkin read Standard Operating Procedure 046, Non-Emergency Incident Directives (Direct Orders) verbatim to you

- You admitted you completely understood the direct order and the SOP that was read verbatim by Chief Buffkin and even initialed the policy when she was done

- Even after Chief Buffkin gave you a direct order (that you completely understood) while at station 30 to issue the reprimands, you still refused to comply with the order

Based on the information listed above, it is clear you refused to comply with a direct order issued by Chief Buffkin on October 5, 2021. Upon evaluation of the information presented, I find you have violated the following:

- Orange County Fire Rescue Department Rules, R.5 Performance of Duty

- Orange County Fire Rescue Department Rules, R.24 Unbecoming Conduct

- Orange County Fire Rescue Department Rules, R.32 Insubordination

- Orange County Fire Rescue Department Rules, R.36 Adhering to and Reporting Improper Orders

- Orange County Fire Rescue Standard Operating Procedures, SOP 046 Non-Emergency Incident Directives (Direct Orders)

(cont. on next page)

- Orange County Firefighters Association Collective Bargaining Agreement (B Unit), Article 12 (1), Grievance and Arbitration Procedures

- Orange County Firefighters Association Collective Bargaining Agreement (B Unit), Article 3, Managerial Responsibilities/Conflicts of Interest

I have reviewed the violations under the Orange County Fire Rescue Department Standard Operating Procedures SOP 06, Corrective Disciplinary Matrix as well as the Orange County Policy Manual & Operational Regulations 413, Disciplinary Action and have determined your actions to be:

1. B. Performance Deficiencies

    (1) Failure to follow operational procedure

2. C. Behavioral Issues/Unsatisfactory Personal Habits

    (4) Dereliction of duty – willful refusal to perform assigned duties or follow a direct order

    (7) Conduct in a work or non-work environment which discredits the County government, public officials, fellow employee(s), or themselves

After consideration of all the information and testimony presented, reviewing the violations of policies, Union Contract Article 10, Discipline, taking into consideration the seriousness of the offense, your employment history, the lapse of time between discipline and the County practice in similar cases, it is my decision to impose the following:

- Your employment with Orange County Fire Department will be terminated at 1700 on October 19, 2021

- You are directed to return all County/Department issued items in accordance with SOP 027, Separation Process to the Supply Bureau located at 400 S. Gaston Foster Road by October 25, 2021

Employees are also encouraged to contact the County's EAP provider, ComPsych whenever they are experiencing personal and/or work related difficulties.

See below for more information on ComPsych®:

Employee Assistance Program (EAP)
ComPsych® GuidanceResources® Program
Call: 855.221.8925
TDD: 800.697.0353
Go Online: guidanceresources.com
Your company Web ID: ORANGECOUNTY

Should you not agree with this decision, you have the right to appeal in accordance with the agreement between Orange County and the Orange County Professional Firefighters Local 2057, Article 12 Grievance and Arbitration Procedures. For the ability to dispute these findings by way of the grievance procedure, your date of knowledge will begin upon receipt of this letter.

Respectfully,

Michael A H

Michael A. Howe
Assistant Chief, Training

c: James M. Fitzgerald, Fire Chief
     David Rathbun, Deputy Chief
     Mike Wajda, Division Chief, Operations
     Kimberly Buffkin, Assistant Chief, C-Shift
     David Hepker, Program Administrator
     Nichol Stratman, Battalion Chief, Centralized Staffing
     Charles Welch, Senior Human Resource Advisor
     Suzette Shields, Compliance and Employee/Labor Relations Administrator
     Ricardo Daye, Director, Orange County Human Resources

ORANGE COUNTY FIRE RESCUE DEPARTMENT
David A. Rathbun, Deputy Fire Chief
P.O. Box 5879
Winter Park, Florida 32793-5879
407-836-9019 Fax: 407-836-9108
David.Rathbun@ocfl.net

Received: _____ UT1 Date: 12/14/21

Delivered: _____ Date 12/14/2021

December 13, 2021

Andre Perez, President
Orange County Fire Fighters Association
6969 Venture Circle
Orlando Florida 32807

Re: Grievance Hearing Decision, Step-2, # 21-12193 (Stephen Davis)

Dear Andre,

A Step-2 grievance hearing was held on November 30, 2021 to address grievance 21-12193. By way of your grievance and verbal testimony, you have alleged that management violated:

- Collective Bargaining Agreement:
  - Articles 2, 3, 6, 9, 10, 12, 16, and 25
- Orange County Fire Rescue Standard Operating Procedures
  - R-2, R-3, R-8, R-33, R-35, SOP-006, and SOP-046
- Orange County Policies:
  - Code of Ethics, 402, and 413

After much deliberation and careful consideration of all written and verbal evidence presented during the hearing, I find that management did not violate the collective bargaining unit articles cited in your grievance. Therefore, your grievance is denied.

Sincerely,

David A. Rathbun
Deputy Fire Chief

C: James M. Fitzgerald, Fire Chief
        Michael Wajda, Division Chief, Operations
        Mike Howe, Assistant Chief
        Charles Welch, Human Resources Senior Advisor

Chief Davis's Step 2 Grievance decision, which went against him.

A Spoliation of Evidence/Public Documents?

> **From:** Izzo, Robert A <Robert.Izzo@ocfl.net>
> **Sent:** Tuesday, October 12, 2021 8:59 PM
> **To:** Deniston, Ben D <Ben.Deniston@ocfl.net>;
> White, Jason G <Jason.White@ocfl.net>; Rhoden,
> Michael A <Michael.Rhoden@ocfl.net>; Roane,
> Sara M <Sara.Roane@ocfl.net>
> **Cc:** Mack, Martis M <Martis.Mack@ocfl.net>
> **Subject:** covid
>
> To all, I have been advised by the AOC that you
> have submitted the required paperwork to HR and
> your written reprimands are not being issued and
> are destroyed. Contact me if you have any
> questions.
>
> BC Robert A. Izzo
> Orange County Fire Rescue
> Battalion 1-A Station 41
> 407-254-8441
>
>
> PLEASE NOTE: Florida has a very broad
> public records law (F. S. 119).
> All e-mails to and from County Officials are
> kept as a public record.
> Your e-mail communications, including your
> e-mail address may be
> disclosed to the public and media at any
> time.

Email from Battalion Chief Robert Izzo stating that erroneously issued reprimands will not be filed and will be destroyed.

Other

HUMAN RESOURCES DIVISION
Orange County Government | Internal Operations Center (IOC-1)
450 East South Street | Orlando, Florida 32801
PHONE (407) 836-5661 • FAX (407) 836-5805

October 29, 2021

Sent via email: Sara.Roane@ocfl.net

Sara Roane
P.O. Box 5879
Winter Park, FL 32793

Subject: Response to Accommodation Request

Dear Ms. Roane:

This letter is to inform you that your religious accommodation request received on September 30, 2021 has been **denied**.

The County reviewed your request, along with any supporting documentation or discussion that accompanied or followed it, and found that it did not rise to the level of a sincerely held religious belief as contemplated by the law, or it would create an undue hardship for the County.

Please know the Employee Assistance Program (EAP) is available to employees, if needed. The Employee Assistance Program (EAP) can be contacted at:
   ComPsych® Guidance Resources® Program
   Call: 855.221.8925
   TDD: 800.697.0353
   Go Online: guidanceresources.com
   Your company Web ID: ORANGECOUNTY

If you have questions, please feel free to contact me at (407) 836-0088.

Sincerely,

Charles Welch

Charles Welch
Sr. Human Resources Advisor
Human Resources Division

cc: James Fitzgerald; Fire Chief, Fire Rescue
       Michael Wajda; Division Chief, Fire Operations
       Ricardo Daye; Human Resources Director
       Suzette Shields; HR Compliance & Labor/Employee Relations Administrator
       File

Email from Orange County to an employee denying their religious accommodation request on account of a lack of 'sincerity.'

The Orange County MOU document which outlines a policy for terminating an employee reg. Covid-19 vaccinations. Note #5 of page 1 has been crossed out. This document was never officially published to personnel, which violated Orange County Policy and the Fire Union's collective bargaining agreement.

County IAFF A & B-Unit Proposal 9-30-21

COUNTY COVID-19 MANDATORY VACCINATION
POLICY AGREEMENT

To address the impacts of the changes to the County Safety Manual implementing the County's mandatory COVID-19 vaccination policy, as reflected in the County's July 30, 2021 correspondence to the Union, the County and the Union agrees to the following:

1. The deadlines for receiving the COVID-19 vaccination have been extended as follows. Bargaining unit employees shall be required to receive the Johnson & Johnson vaccine, or the first Pfizer or Moderna vaccine, and complete and submit their COVID-19 Vaccination Certification Form by September 30, 2021. Bargaining unit employees shall be required to receive the second dose of the Pfizer or Moderna vaccine, and complete and submit their COVID-19 Vaccination Certification Form by October 31, 2021.

2. Bargaining unit employees who receive the Johnson & Johnson vaccine, or the first dose of the Pfizer or Moderna vaccine, and complete and submit their COVID-19 Vaccination Certification Form on or before September 30, 2021, shall be provided one (1) full shift off with pay. The shift off with pay must be taken by October 31, 2022, will not be paid out on separation from employment, and is subject to supervisor approval. Additionally, bargaining unit employees who received the Johnson & Johnson Vaccine, or the first does of the Pfizer or Moderna Vaccine, and submitted their COVID-19 vaccination certification on or before August 31, 2021, shall receive a $250 Lump Sum incentive paid in October 2021 or as soon thereafter as possible.

3. Bargaining unit members shall be allowed to attain COVID-19 vaccinations, including all County required boosters, while on duty.

4. Bargaining unit members may request a Medical or Religious accommodation/exemption to the COVID-19 vaccination. Accommodation requests must be submitted on or before September 30, 2021. Members who are granted an exemption and remain unvaccinated shall be subject to required weekly COVID-19 testing conducted in accordance with the OCFR G.O. for Testing Procedures, and other safety measures determined by management (e.g., masks while on duty).

5. ~~Bargaining unit members who, as of the deadlines remain unvaccinated or have not submitted an accommodation request and supporting documentation, and bargaining unit members who do not submit certification of vaccination within five (5) business days of the denial of their accommodation request, shall be subject to a written reprimand. COVID-19 testing in accordance with the OCFR G.O. for Testing Procedures, and other safety measures determined by management (e.g., masks while on duty). No further disciplinary action shall be taken for failing to comply with the COVID-19 vaccination policy deadlines.~~

Bargaining unit members who do not complete and submit their COVID-19 Vaccination Certification Forms within the deadlines, who do not complete and submit their accommodation requests and supporting documentation within the deadlines, or who do not complete and submit their COVID-19 Vaccination Certification Forms within ten (10) business days of the denial of their accommodation requests, shall be subject to a written reprimand. No further disciplinary action shall be taken for failing to comply with the COVID-19 vaccination policy deadlines. This written reprimand shall not be considered or used in the bargaining unit member's performance evaluation. Additionally, such

1 of 2

bargaining unit employees shall be required to undergo weekly COVID-19 testing conducted in accordance with the OCFR G.O. for Testing Procedures, and other safety measures determined by management (e.g., masks while on duty).

6. Bargaining unit members who fail to comply with required weekly COVID-19 testing and other safety measures determined by management (e.g., masks while on duty) shall be subject to disciplinary action as provided in the collective bargaining agreement and applicable policies, and shall not be limited to a written reprimand.

7. This Agreement shall be non-precedent setting, and shall not be used as an interpretation of the A and B unit Collective Bargaining Agreements.

8. The terms herein shall go into effect upon the signing of this Agreement and shall continue through the end of the County's officially declared State of Emergency, or upon mutual agreement, whichever is earlier, unless reactivated by a subsequent County officially declared State of Emergency regarding COVID-19.

9. Unless specified herein, the terms and conditions of the A and B unit Collective Bargaining Agreements will remain intact and unchanged.

10. By entering into this Agreement, neither party waives its positions regarding bargaining over the County's mandatory COVID-19 vaccination policy, and the Union does not waive any right to bargain over future vaccination issues.

FOR THE COUNTY:

Jeffrey Mandel
County Labor Counsel

9/30/21
(Date)

FOR IAFF LOCAL 2057:

Andre Perez
President

9/30/21
(Date)

- Commercial Testimonials

  Employees shall not permit their names or photographs to be used in endorsing any product that is service-connected with the Fire and Rescue Department without the permission of the Fire Chief, and shall not allow their names or photographs to be used in any commercial testimonial, which alludes to their positions or employment with the Department.

- Personal Publicity

  Employees shall not use their positions within the Department to enhance or promote any private enterprise, or to seek personal publicity.

- Public Appearances

  Requests for public speeches, presentations, and the like shall be routed to the affected Division Chief for approval and processing. Employees directly approached with requests of this nature shall refer the party to their division supervisors.

- Publishing Articles

  Employees, as authors or co-authors of any articles representing their positions in the Department or the policies of the Fire and Rescue Department, shall submit the article to the Fire Chief and the Public Information Office for review and approval, prior to submission of the article for publication.

R.32 Insubordination

Except as otherwise stated herein, defiance of a superior officer or disobedience to orders constitutes insubordination.

R.33 Unlawful Orders

Employees shall not knowingly issue any order that is in violation of laws, statutes, ordinances, rules and regulations, or SOPs.

R.34 Manner of Issuing Orders

Orders shall be issued in a clear and civil tone, in an understandable manner, and in the interest of Department business.

R.35 Disobedience to Unlawful Orders

No employee is expected to, or shall obey any order, that he/she knows to be contrary to federal or state law or County ordinance. At the time an unlawful order is issued, the employee shall:

- Advise the issuing authority of its illegality.

- Should that authority persist in demanding compliance, an employee of superior rank or status to all parties involved should be summoned to decide the controversy.

Rules of Orange County Fire Department. Rules 33 and 35 have been highlighted.

List of names of OC personnel that were to be reprimanded over the Covid-19 policy. This is only the C-Shift list and does not include A and B shifts and other employees. Highlights in green were listed for reprimand improperly. Note the blue-highlighted section at the bottom were all members of Davis's battalion of Fire and EMS personnel, with two improperly listed individuals.

112480	Helms,Erik Joseph	1143	EN	Operations	1	27	C
130218	Davalos,Jade Justin	2456	FF/PM	Operations	1	28	C
131485	Johnson,Joseph Matth	2542	FF/PM	Operations	1	28	C
133180	Manuel,Austin David	2837	FF	Operations	1	28	C
132180	Souffront,Kelvin Luis	2596	FF/PM	Operations	1	28	C
132802	Joseph,Ronald	2741	FF	Operations	1	40	C
116493	Cezalien,Billy	1382	EN	Operations	1	41	C
130906	Greer,Ronald Donel	2500	FF	Operations	1	41	C
125611	Sledge,Jared D.	2024	FF/PM	Operations	1	41	C
117493	Green Jr.,Dwight E.	1427	EN	Operations	2	30	C
104997	Hawkins,Jack E	867	FF/PM	Operations	2	30	C
122573	Pierre,Wesley	1819	EN	Operations	2	30	C
132161	Harris Diaz,Edrick Jam	2605	FF	Operations	2	31	C
120024	De Jesus,Richard P.	1599	FF	Operations	2	33	C
132186	Garcia,Isaiah Andraes	2578	FF	Operations	2	33	C
129480	Soto Sierra,Alberto Or	2347	FF/PM	Operations	2	33	C
116496	Stephens,Daris J.	1388	EN	Operations	2	33	C
133337	Alectus,Kervens	2880	FF	Operations	2	42	C
132803	Hamilton,Nathan R	2739	FF	Operations	2	43	C
104935	Kleiman,Stephen L	871	LT/PM	Operations	2	43	C
123681	Wright,Ian-Michael	1887	EN	Operations	2	43	C
127498	Coll Ramos,Christ Ang	2134	FF/PM	Operations	2	52	C
133108	Marquez Betancourt,E	2807	FF	Operations	2	52	C
111099	Chaples,Donald Willia	1011	FF/PM	Operations	3	54	C
125618	Moody,Clayton Earle	2018	FF/PM	Operations	3	54	C
129136	Young,James Tucker	2273	FF/PM	Operations	3	54	C
129442	Castillo,Christopher Cl	2316	FF/PM	Operations	3	55	C
129762	Wildey,Chelsea Rae	2401	FF/PM	Operations	3	55	C
122212	Anhalt,John Robert	1785	EN/PM	Operations	3	58	C
129471	Nissen,John Elijah	2346	FF/PM	Operations	3	58	C
130211	Spencer,Charles Lawre	2418	FF	Operations	3	58	C
122570	Martinez III,Sotero	1816	EN/PM	Operations	3	76	C
105259	Santiago,Alex	977	LT/PM	Training	3	76	C
132136	Cuevas,Vicente Cecilio	2572	FF	Operations	3	77	C
107835	Hart,Daniel P	922	EN/PM	Operations	3	77	C
121756	Brett,Daniel Allen	1743	EN/PM	Operations	4	50	C
120534	Hansen,James W.	1629	LT/PM	Operations	4	50	C
129134	Ickes,Brandon Scott	2282	FF/PM	Operations	4	50	C
132189	Silverman,Luke Alan	2595	FF	Operations	4	50	C
129769	Willard Bennett,Sara N	2402	FF/PM	Operations	4	51	C
132564	Brodrecht,Ethan Jame	2615	FF	Operations	4	70	C
117573	Coats,Daniel B	1448	LT	Operations	4	70	C
133811	Mesa,Evelio E	2932	FF	Operations	4	70	C
132143	Torrez,William Anthon	2597	FF	Operations	4	70	C
121169	Ping,Phillip Thomas	1699	EN	Operations	4	72	C
130216	Salafrio,Nickolas Lowe	2464	FF/PM	Operations	4	72	C
110202	Lambis,Roberto C	1440	FF	Operations	4	73	C
133552	Gonzalez,Jr.,Jorge L	2910	FF	Operations	5	63	C
126045	Sauerbrun,Matthew N	2042	EN/PM	Operations	5	63	C

1

Printed in the USA
CPSIA information can be obtained
at www.ICGtesting.com
LVHW040529030624
781894LV00007B/722